Medical School Statistics Rx: Statistics for the MCCEE and USMLE

Published by
MCCEE Tutoring Services
841 Brentwood Drive
Port Elgin, ON
N0H 2C4

www.mcceetutoringservices.com

Copyright © 2019 by MCCEE Tutoring Services, Port Elgin, Ontario

Published by MCCEE Tutoring Services, Port Elgin, Ontario

No part of this publication may be reproduced by any means, stored in a retrieval system or transmitted in any form, mechanical, electronic, photocopying, recording, scanning, or otherwise except under the pertaining sections of the Copyright Act, without either the written authorization of the Publisher. Requests to the publisher should be addressed to MCCEE Tutoring Services, 841 Brentwood Drive Port Elgin, ON, N0H 2C4 or online at http://www.mcceetutoringservices.com/contact.

Limit of Liability: The author and the publisher make no representations or warranties with respect to the accuracy or completeness of this work and specifically disclaim all warranties, including without limitation warranties of fitness for a particular purpose. No warranty may be created or extended by sales or promotional materials. The content and strategies contained here may or may not be on your particular exam. This work is sold as a supplement to the learning provided in medical school and adequate preparation for the standardized medical exams. Neither the publisher nor the author shall be liable for damages arising here form. Please remember that Internet Websites listed in this book may no longer exist or changed after this book was written and when it is read.

Preface

Medical School Statistics Rx is a book that intends on teaching statistics in a way that will actually help medical students remember the concepts needed to pass the standardized exams in the least painful way possible. This includes the USMLE Step 1, Step 2 CK, and MCCEE (now a defunct exam). The other goal is to make sure all this knowledge you gain can be applied later on when you have to judge a paper by the strength of its statistical arguments.

For those of you who want a dry book on statistics that makes you rethink your medical career this will not be the book for you! The book assumes that you, as the reader, will work on the exercises that have been made available. I am sure that by actively studying the material in this way as opposed to passively studying will help you assimilate the statistics that your professors are forcing you to learn.

The first chapter (Chapter 1) goes over the basic definitions you will need to know and some very minor calculations. The following chapter (Chapter 2) deals with the standard calculations you will face on exams. The third chapter (Chapter 3) encompasses the study types, such as cohort, and how these affect what calculations can be done as well as dealing with the many different study biases you will come across on exams and in the real world. The final chapter (Chapter 4) will show you examples of questions on the USMLE exams. Although these questions will not replicate the exact way questions will be asked, they will teach you the skills to go through a journal article problem. These problems typically encompass you looking at a journal article and answer questions about it.

These chapters will attempt to go through the high yield objects of the MCCEE and MCCQE shown here, but there is a great amount of overlap with the USMLE exams.

As a disclaimer this book should be used as a supplement to your studying and assists with understanding. It is not necessarily all encompassing and the questions are meant to aide with understanding and not simulate the exams perfectly.

78-2
December 2012

Assessing and Measuring Health Status at the Population Level
Population Health

Rationale

Knowing the health status of the population allows for better planning and evaluation of health programs and tailoring interventions to meet patient/community needs. Physicians are also active participants in disease surveillance programs, encouraging them to address health needs in the population and not merely health demands.

Key Objectives

- Describe the health status of a defined population.
- Measure and record the factors that affect the health status of a population with respect to the principles of causation.

Enabling Objectives

- Know how to access and collect health information to describe the health of a population:
  - Describe the types of data and common components (both qualitative and quantitative) used in creating a community health needs assessment.
  - Be aware of important sources of clinical / population-level health data and recognise the advantages and disadvantages of each of them.
  - Critically evaluate possible sources of data to describe the health of a population including the importance of accurate coding and recording of health information.
  - Describe the uncertainty associated with capturing data on the number of events and populations at risk.
  - Understand surveillance systems and the role of physicians and public health in reporting and responding to disease.
- Analyze population health data using appropriate measures:
  - Apply the principles of epidemiology in analyzing common office and community health situations.
  - Describe the concepts of, and be able to calculate, incidence, prevalence, attack rates, case fatality rates and to understand the principles of standardization.
  - Discuss different measures of association including relative risk, odds ratios, attributable risk and correlations.
- Interpret and present the analysis of health status indicators:
  - Demonstrate an ability to use practice-based health information systems to monitor the health of patients and to identify unmet health needs.
  - Understand the appropriate use of different graphical presentations of data.
  - Describe criteria for assessing causation.
  - Demonstrate an ability to critically appraise and incorporate research findings with particular reference to the following elements:
    - characteristics of study designs (RCT, cohort, case-control, cross sectional);
    - measurement issues (validity, sensitivity, specificity, positive predictive value, negative predictive value; bias, confounding; error, reliability);
    - measures of health and disease (incidence and prevalence rates, distributions; measures of central tendency) and sampling.

If you need more information or tutoring for your exams, shelf, MCCEE, USMLE, and other, please send me a message at the email mccee.tutor@gmail.com or leave a note on the website at http://www.mcceetutoringservices.com/contact. My team and I will be glad to assist you. All initial consults are no obligation and free of charge.

<u>Acknowledgements</u>

This book could not have been completed without the help of a couple of people. I would like to thank Gamal Fitzpatrick for his continued support of the tutoring service as a whole and for taking the time to edit the book. A special thanks goes to Nina Chan who helped look at the book from a student perspective and offered great advice. To my family who have continued to support me through my medical career. I also could not have had the inspiration to create this book if it was not for Lindsay Abraham whose love and support has been a driving factor for success in my medical career. Thank you all for the assistance!

Table of Contents

Introduction

How to use this text:

1. Read the interactive text
   a. Fill in the blanks using the answers on each page
   b. Use page illustrations to help with understanding
2. Do not turn to the next page until after you have fully understood the concept
3. Use the review questions to help further understand and remember each concept
   a. Some questions will be multiple choice questions (MCQ) that will help to *simulate* the style you will find on your standardized exams. The actual exam questions may be harder or easier than the ones here
   b. Other practice questions will be more short answer format which more strongly assesses your understanding
4. At least pretend to enjoy the math!

Sincerely,

Brock Juffs
Founder of MCCEE Tutoring Services

Chapter 1: Definitions

Prevention Strategies

Type of prevention	Description	Case
Primary	Prevent disease from developing	Smoking prevention campaigns
Secondary	Early detection of an existing disease with the aim of reducing complications and severity	Screening mammography
Tertiary	Decrease the effect of the disease	HIV medications

Part of epidemiology is preventing diseases from ever occurring or progressing. There are three types of prevention.

_____ _____ is preventing a disease from occurring to begin with. An example is _____. Not 100% of the population needs to be immunized for all disease due to ____ _____.

Primary prevention, immunization, herd immunity

Secondary prevention entails _____ _____ of existing disease to reduce severity and existing complications, potentially catching a disease in its _____ stage. An example is _____ _____.

early detection, cancer screening, preclinical

Tertiary prevention aims to reduce the _____ of disease. An example is _____ after a myocardial infarction.

impact, rehabilitation

Triad of Disease

Epidemiological factors affecting disease are classically shown in a triad.

One characteristic of disease transmission is the _____. Examples of these characteristic are age and race.

host

The _____ of disease transmission refers to the direct causative factor such as bacteria.

agent

The environment can be something such as a _____ water supply that leads to legionella infection.

contaminated, Legionella

Endemic, Epidemic, and Pandemic

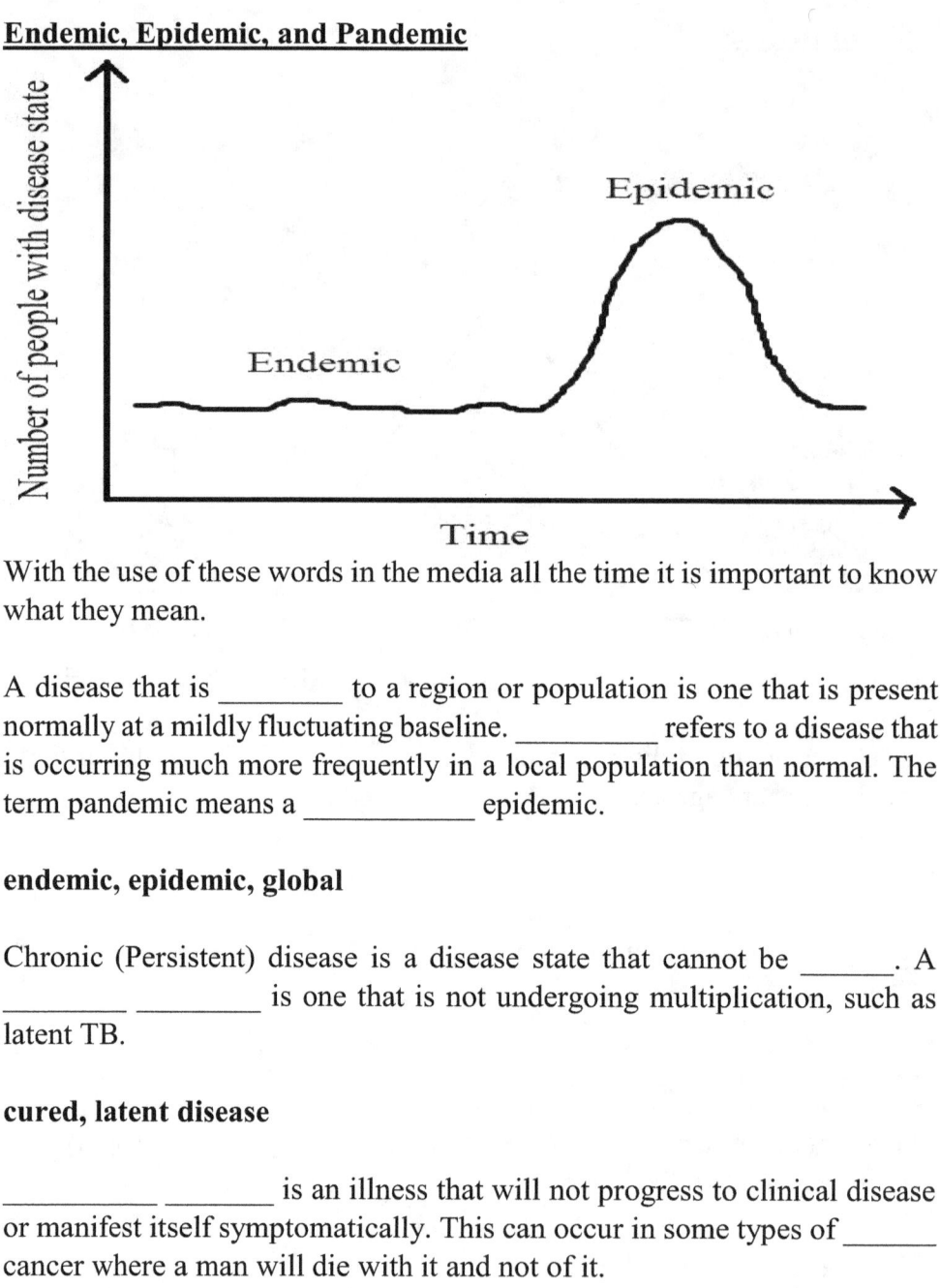

With the use of these words in the media all the time it is important to know what they mean.

A disease that is _____ to a region or population is one that is present normally at a mildly fluctuating baseline. _____ refers to a disease that is occurring much more frequently in a local population than normal. The term pandemic means a _____ epidemic.

endemic, epidemic, global

Chronic (Persistent) disease is a disease state that cannot be _____. A _____ _____ is one that is not undergoing multiplication, such as latent TB.

cured, latent disease

_____ _____ is an illness that will not progress to clinical disease or manifest itself symptomatically. This can occur in some types of _____ cancer where a man will die with it and not of it.

Subclinical disease, prostate

Dataset Types

Ordinal   Interval   Ratio

Datasets have multiple types. Nominal (categorical), continuous, discrete, ratio, interval, and ordinal data.

_____ data is a set of values that have an order (ORDer and ORDinal). An example would be the ranks of _____ athletes.

Ordinal, Olympic

Interval data is a set where intervals between each value are _____ split. An example is _____.

equally, temperature

_____ data is categorical without defined values. An example would be _____ or the ABO blood groups.

Nominal, gender

Continuous data means any data point is _____. _____ data differs from continuous data in that not all data points are useful. An example is the number of children in a household cannot be anything other than a _____ number.

meaningful, Discrete, whole

Note: This information may seem elementary but it is important to know the data set type when doing an analysis on it. We will come back to this later.

Median, Mean, and Mode

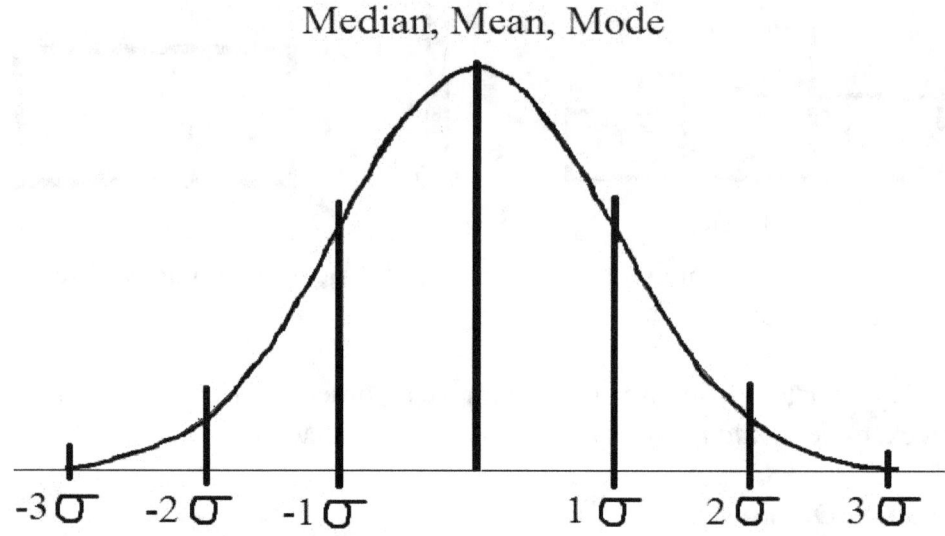

There are three measures of central tendency. The median is always the middle value of the set by definition (splits the dataset in two). The mean is an average and the mode is the most common number.

The _____ is the middle value of a data set. It is _____ affected by extremes (including outliers).

median, least

The _____ is the most affected by extremes (including outliers). It is calculated by the following formula: $m = \frac{\sum_{i=1}^{n} x_n}{n}$ where m is the mean, n is the set size, and x denotes a specific value in the set. The mode is the value that _____ the most and is not affected by outliers.

mean, recurs

In a standard _____ _____ as above the mean, median, and mode are all equal. This is known as _____ distribution.

bell curve, Gaussian

Note: The symbol in the above graphic is *sigma* and it denotes a standard deviation, which will be discussed later.

Skew

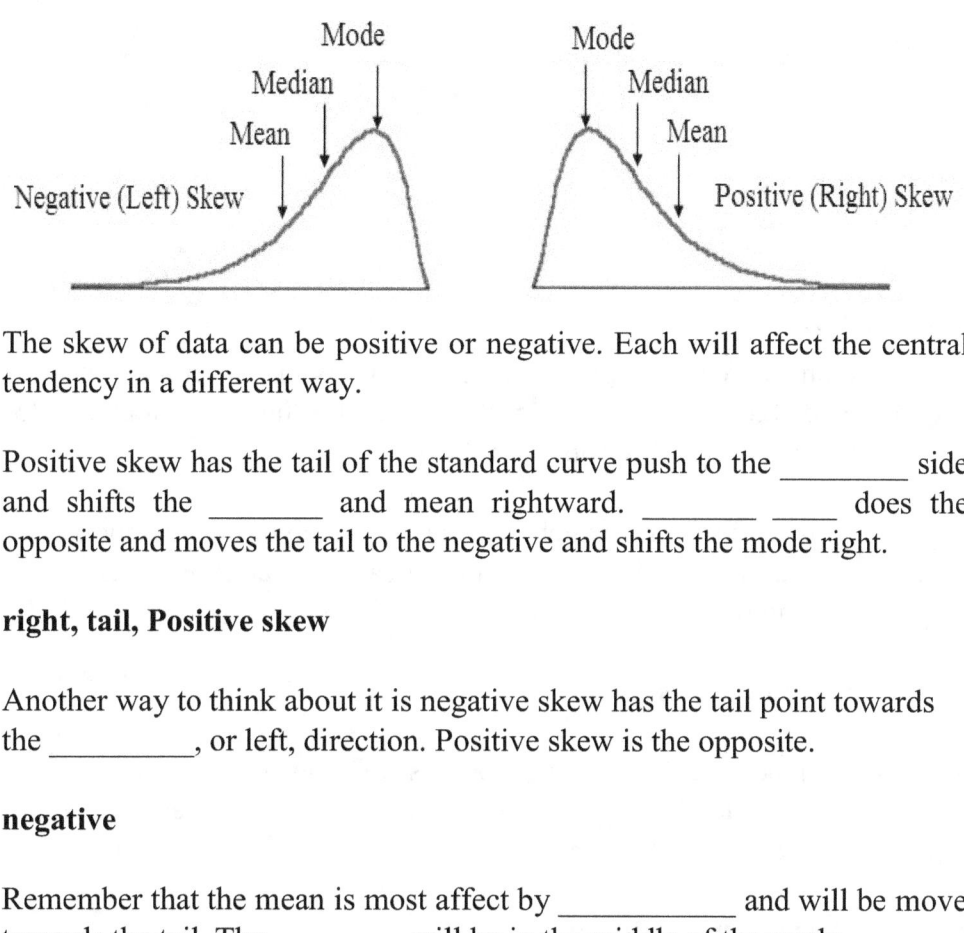

The skew of data can be positive or negative. Each will affect the central tendency in a different way.

Positive skew has the tail of the standard curve push to the _____ side and shifts the _____ and mean rightward. _____ _____ does the opposite and moves the tail to the negative and shifts the mode right.

right, tail, Positive skew

Another way to think about it is negative skew has the tail point towards the _____, or left, direction. Positive skew is the opposite.

negative

Remember that the mean is most affect by _____ and will be move towards the tail. The _____ will be in the middle of the mode.

extremes, mode

Accuracy and Precision

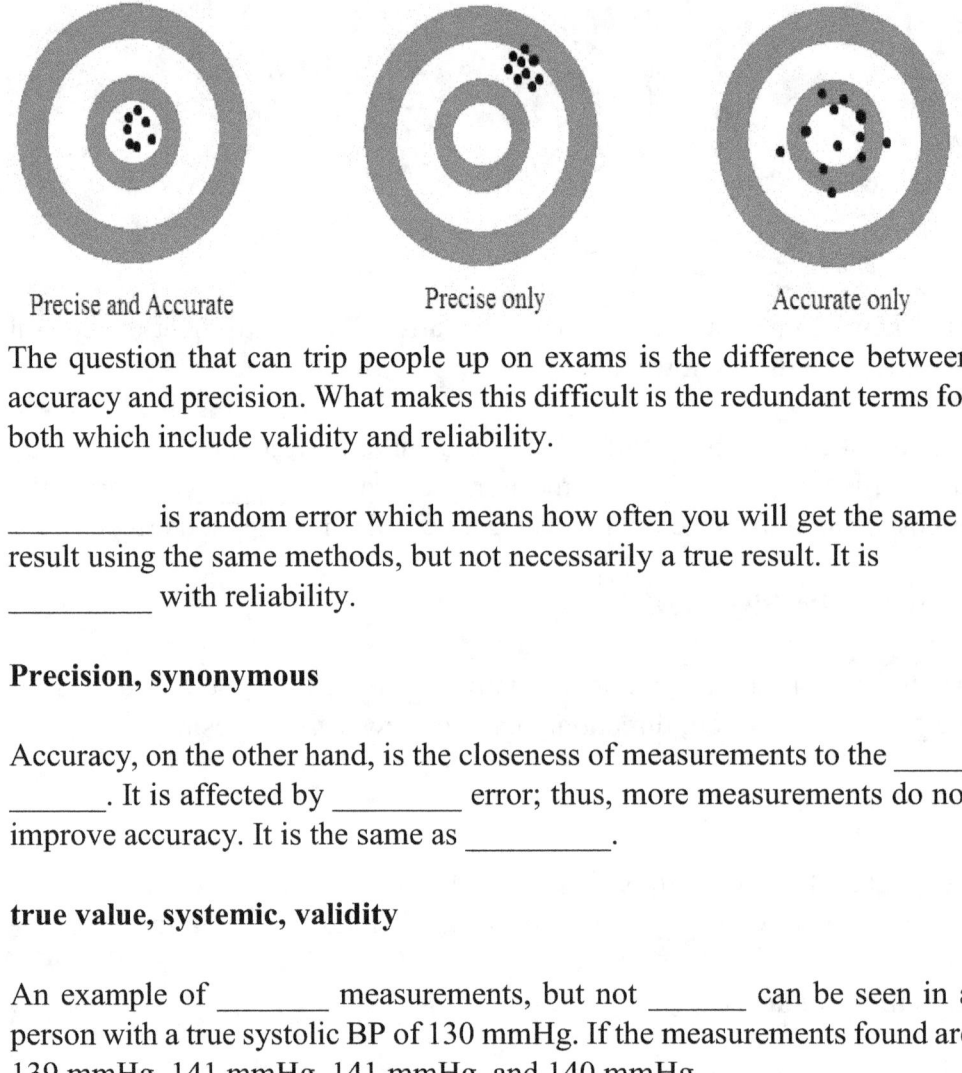

Precise and Accurate   Precise only   Accurate only

The question that can trip people up on exams is the difference between accuracy and precision. What makes this difficult is the redundant terms for both which include validity and reliability.

_____ is random error which means how often you will get the same result using the same methods, but not necessarily a true result. It is _____ with reliability.

Precision, synonymous

Accuracy, on the other hand, is the closeness of measurements to the _____ _____. It is affected by _____ error; thus, more measurements do not improve accuracy. It is the same as _____.

true value, systemic, validity

An example of _____ measurements, but not _____ can be seen in a person with a true systolic BP of 130 mmHg. If the measurements found are 139 mmHg, 141 mmHg, 141 mmHg, and 140 mmHg.

precise, accurate

Internal _____ is affected by biases, low study _____, and multiple tests of significance. All of these will be discussed in later in this text. _____ validity is affected by the study population and its _____ to the target population.

Validity, power, external closeness

Intrasubject Variation

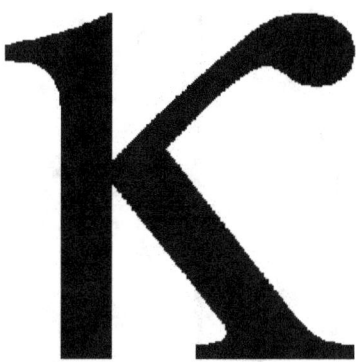

Kappa

An ideal exam finding is one that physicians can agree upon more often than chance alone would expect. Kappa measures this.

$$\kappa = \frac{x - y}{100\% - y}$$

Where **x** is the **actual percent agreement**, and **y** is the **expected percent agreement by chance alone**

The higher the value of kappa the _____ likely that the exam finding will be _____ consistently observed between physicians.

more

_____ variation is variation between the interpretation of a test done by the same individual. An example is the interpretation of the same CT scan at different times and coming up with a different findings each time.

Intrasubject

Intraobserver variation is when ____ examiners do not come up with the same result. This can be seen with medical students who do not agree on the characteristics of the _____ they are examining.

two, rash

Standard Deviation

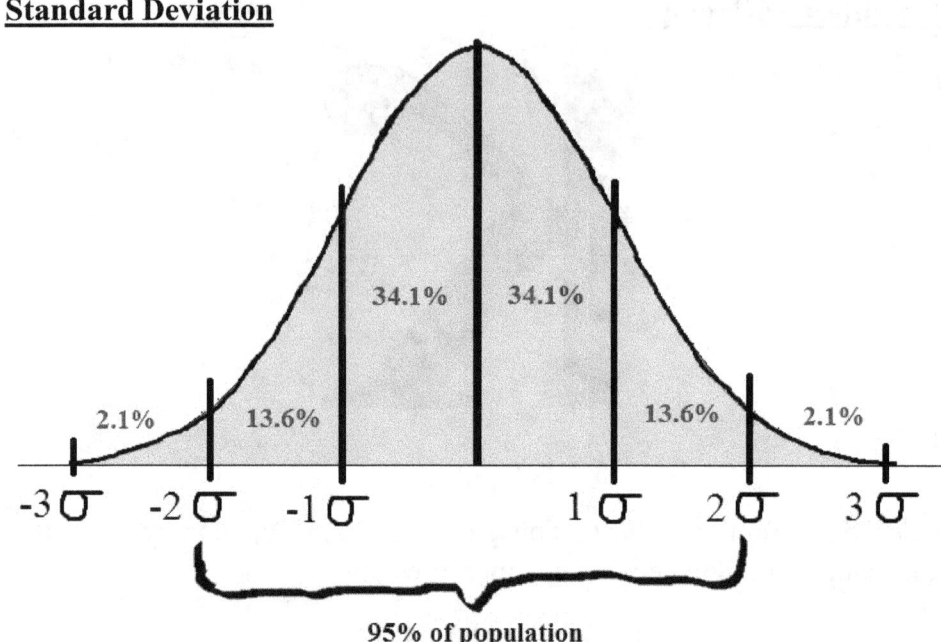

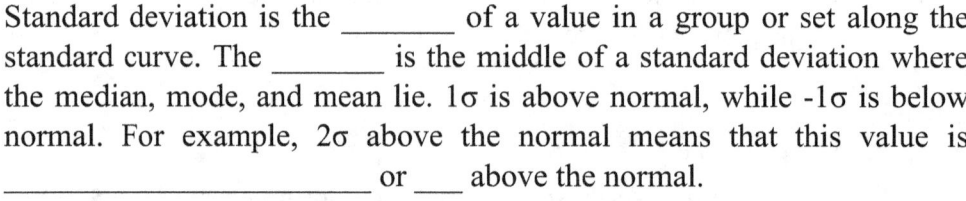

95% of population

A common question on tests is about the standard distribution and standard deviation (SD), σ. Calculating SD is beyond the scope of the exams (it is a lengthy process), but understanding it is not. **You are expected to have this graphic memorized!**

Standard deviation is the _____ of a value in a group or set along the standard curve. The _____ is the middle of a standard deviation where the median, mode, and mean lie. 1σ is above normal, while -1σ is below normal. For example, 2σ above the normal means that this value is _____ or ____ above the normal.

deviation, normal, 2.1 + 13.6 + 34.1 + 34.1 + 13.6, 97.5%

This means if a student scores _____ above the normal on an exam the scored higher than 99.6% of the class.

3 SD

Note: The area shown above the graph does not add up to 100%, but if you were to add more standard deviations it will eventually add to 100%.

Standard Error of the Mean vs Standard Deviation

SEM ≠ Standard Deviation

Try not to get the SD and the standard error of the mean (SEM) confused. By definition the SEM is always smaller than the SD.

The SEM quantifies how _____ you know what the true mean is from your sample data. As shown below, as you sample size _____ the SEM decreases and becomes more precise.

precisely, increases

$$SEM = \frac{\sigma}{\sqrt{n}}, \text{ where n is the sample size and } \sigma \text{ is the SD.}$$

$$\sigma = \sqrt{\frac{\sum_{i=1}^{n}(x_i - \bar{x})^2}{n-1}} \text{ Where n is the number in the set and x-bar is the}$$

average, and Σ is means the sum of.

SD does not change _____ as you gather more data if you look at its complicated equation.

predictably

Note: One of the key features here is to realize that you will never have a sample that perfectly models the population as a whole; however, as your sample gets larger you should get an answer that is closer to the true value

Confidence Interval

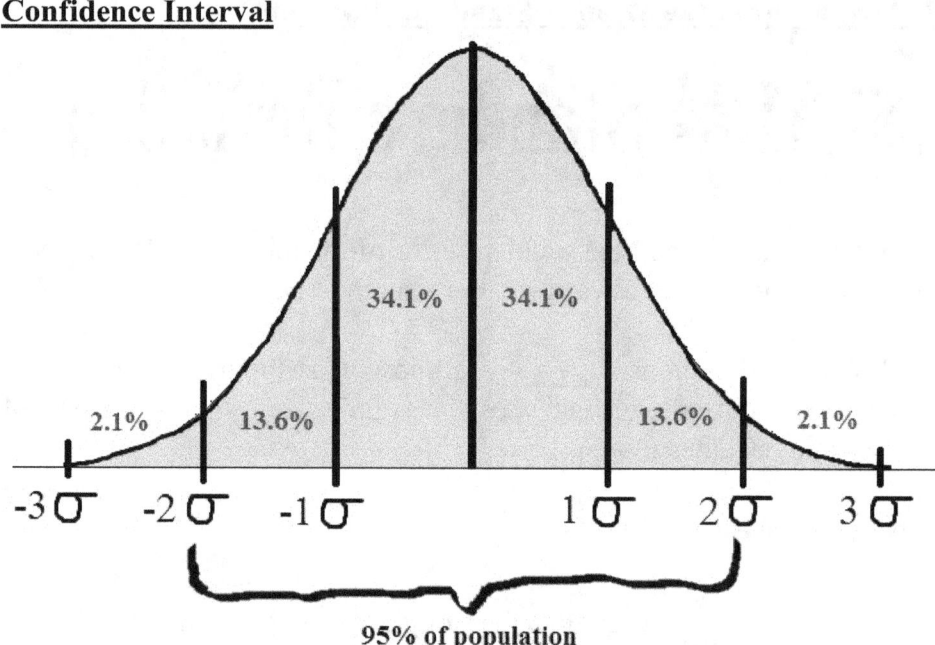

95% of population

Just like a few definitions we will see soon, the confidence interval (CI) is often misunderstood.

A ____ confidence interval is the most commonly seen. It means that 95% of the estimates made will contain the sample parameter (if you took more than one sample); thus, it describes the _____ of the sampling method, not the _____.

95%, uncertainty, certainty

The Z-score (discussed on the following page) can be used is a simple way on USMLE exams to find the margin of error (ME). For the **95% CI the Z-score can be 1.96**, while a **99% CI will have a Z-score of 2.58**. In real life these Z-scores must be calculated.

The ME will be the following: ME = Z * SEM which means the CI = mean +/- ME. This means as you _____ the sample size the CI becomes narrower.

narrower

Z-score

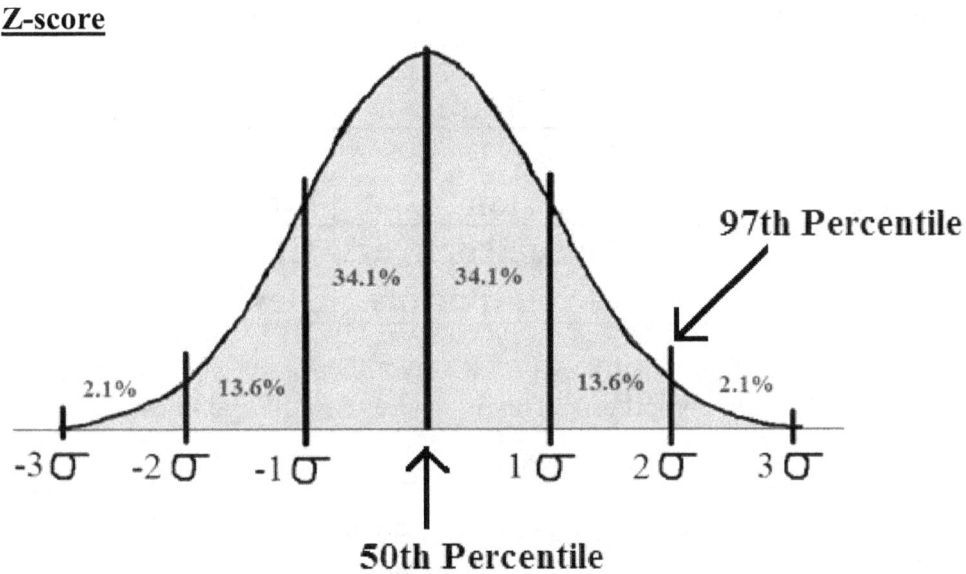

$Z = \frac{x - \bar{x}}{\sigma}$ where x is the raw score, $\bar{x}$ is the mean of the sample, and σ is the SD.

The Z-score _____ how many standard deviations a value in a set is from the normal. A _____ value is below the normal, while _____ is above it.

evaluates, negative, positive

A Z-score can calculate a _____, which is states what portion of the population that value is in. For instance, the ___ _____ means that the value is _____ than half of the values in the set, and ____ than the other half.

percentile, 50ᵗʰ percentile, greater, less

Note: The table in the appendix will help you correlate Z-scores with percentiles

Null Hypothesis

		Actual truth	
		H_1 True	H_0 True
Study Results	H_1 True	**Correct Decision**	False Positive (Type I Error)
	H_0 True	False negative (Type II error)	**Correct Decision**

In science you cannot positively prove a hypothesis, but you can disprove and *fail to disprove* hypothesis. This is where the null hypothesis (H_0) fits in.

The ____ _____ states that there is no relationship between the variables of study. For example no _____ between a toxin and a disease.

null hypothesis, association

The alternate hypothesis (H_1) states that there ____ ___ some difference. The goal of every study is to see if the null hypothesis will be _____ or fail to be. Later chapters will discuss how this is done, and what the exact meaning of the power of a study and what the types of error are are.

may be, rejected

Stratified Data

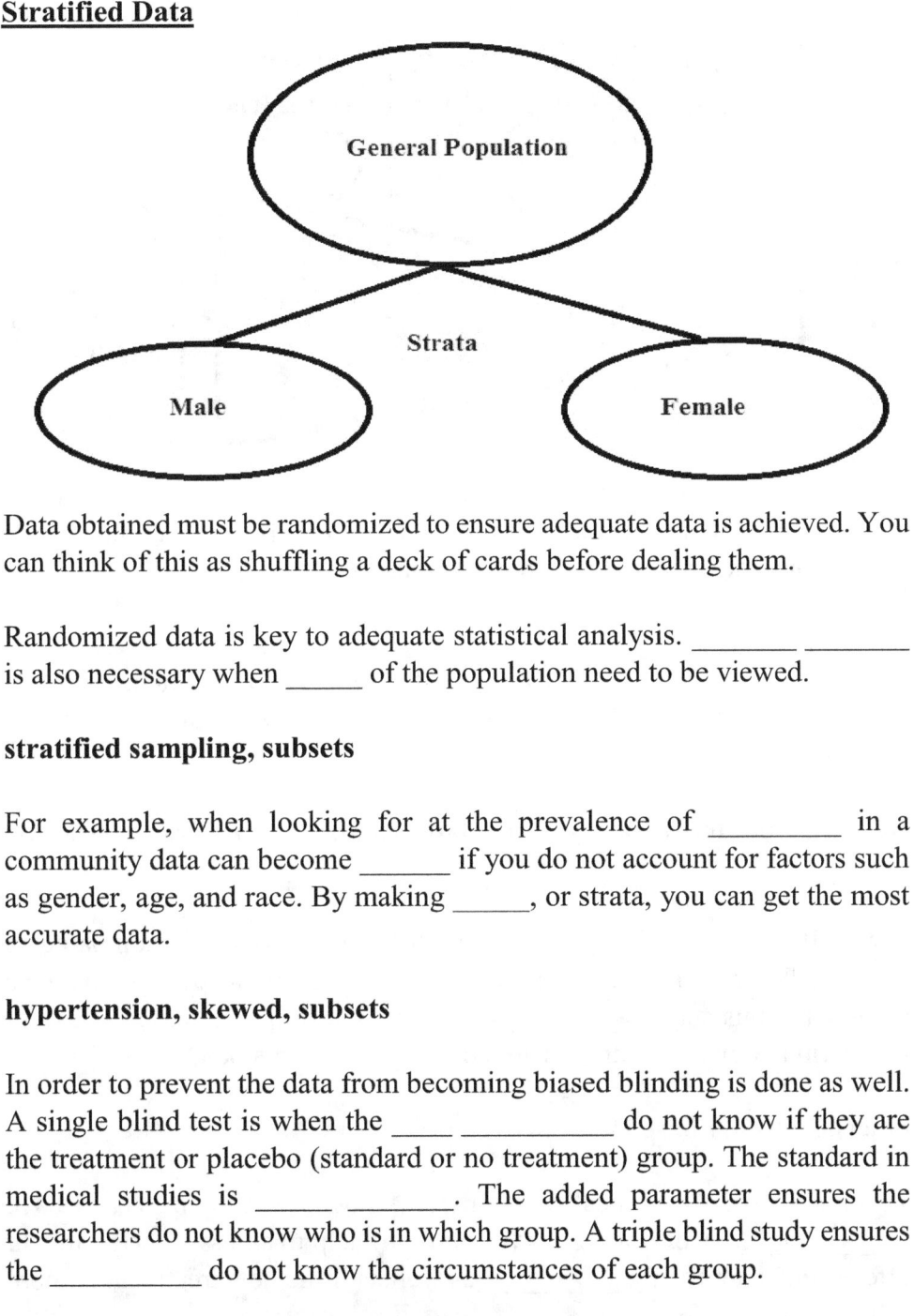

Data obtained must be randomized to ensure adequate data is achieved. You can think of this as shuffling a deck of cards before dealing them.

Randomized data is key to adequate statistical analysis. _____ _____ is also necessary when _____ of the population need to be viewed.

stratified sampling, subsets

For example, when looking for at the prevalence of _____ in a community data can become _____ if you do not account for factors such as gender, age, and race. By making _____, or strata, you can get the most accurate data.

hypertension, skewed, subsets

In order to prevent the data from becoming biased blinding is done as well. A single blind test is when the ____ _____ do not know if they are the treatment or placebo (standard or no treatment) group. The standard in medical studies is _____ _____. The added parameter ensures the researchers do not know who is in which group. A triple blind study ensures the _____ do not know the circumstances of each group.

study participants, double blinding, statisticians

Clinical Trials

Clinical trials are studies that involve human testing. Preclinical trials study animal models. Clinical trials are broken into four phases that increase the number of patients involved with each consecutive phase.

Phase I involves _____ _____. The goal is to assess the drug's _____ and _____. _____ involves the first group of patients with the condition to be studied. This allows _____, safety, and _____to be calculated.

Healthy volunteers, safety and pharmacokinetics, Phase II, efficacy, dosing

Phase III differs from Phase II in that the new treatment is compared to _____, the current treatment, or both. There are also ____ patients involved at this stage. _____ starts after the drug is released to assess long term toxicity. The drug is already _____ at this point.

placebo, more, Phase IV, approved

To be effective the clinical trial must be blinded, randomized, and have _____ _____. Control groups (or placebo) help to assess if the treatment effects are better than _____ _____ as well as if the new treatment is more efficacious.

Adequate controls, chance alone

<u>End of Chapter 1 Review Questions</u>

1. Explain and describe the three different preventive measures of medicine. Provide an example of each.

2. A 71 year old male with COPD, diabetes mellitus, and peripheral is recovering from an acute exacerbation of his COPD. He is having difficulty ambulating without becoming too short of breath. No pedal edema is present or ascites. SaO_2 is 87% but increases to 91% with supplemental oxygen. CXR is negative except for evidence of air trapping. ABG shows O_2 of 53 mmHg. The physician decides to start him on supplemental oxygen for home. What kind of prevention is this?
   a. Primary
   b. Interventional
   c. Tertiary
   d. Invasive
   e. Secondary

3. A 24 year old male is in the ER with complaints of weight loss, night sweats, and shortness of breath. He is an inmate at a local prison. When considering the different diagnoses what triad of disease transmission due you consider?
   a. Host, environment, agent
   b. Exposure, Toxin, Agent
   c. Environment, patient, vector
   d. Environment, agent, vector
   e. Vector, host, environment

4. Explain the difference between endemic, epidemic, and pandemic.

5. What is the difference between the second and third phase of clinical trials?

6. How would you explain what a null hypothesis is to one of your colleagues? How does it relate to the alternate hypothesis?

7. The day finally arrives and you take the MCCEE. It takes weeks but you finally receive your score. You have a 353, with a mean score of 271, and SD of 50. Congratulations! What is your percentile (hint: use the appendix at the end of the book)?
   a. 85th
   b. 95th
   c. 45th

    d. 90th

    e. 99th

8. After years of collecting data you are ready to see how the flu affects different communities. In a particular hospital of 1000 patients and employees 99 get the flu. According to your study per a population of 1000 people 59 would be expected to have the flu. The SEM of your study is 3 on a 95% CI.

    a. What is the margin of error of your study?

    b. Assuming a standard deviation of 30, how does the nursing home compare to the study population in terms of the percentile (do not use the appendix)?

9. You are working on an analysis of a population and you notice the finding in the distribution below.

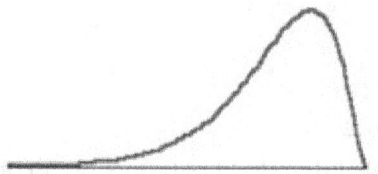

    a. What distribution is this?

        i. Gaussian

        ii. Positive skew

        iii. Standard

        iv. Left skew

        v. Right skew

    b. What is the distribution of the mean, median, and mode?

        i. mode > mean > median

        ii. mode > median > mean

        iii. median > mean > mode

        iv. median > mode > mean

        v. mean > median > mode

10. You receive your last six test scores back and they are 97%, 76%, 27%, 76%, 88% and 78%. Which of the following is true?

    a. The median is smaller than the mode

    b. The mean is larger than the mode

    c. The mean is larger than the median

    d. The mode is larger than the median

    e. The median is larger than the mean

11. A company is trying to see how their lab test for a specific disease compares to the leading test. There test is able to obtain true positive results 97% of the time with a SD of 4. The leading test obtains a true positive 93% of the time with an SD of 2. Compared to the leading test the new test is:
    a. more accurate, but less valid
    b. more precise, but less valid
    c. more accurate, but less reliable
    d. more reliable, but less precise
    e. more precise, but less accurate

12. You notice on an exam that you scored a scaled score of 75 on an exam with a mean of 100 and SD of 25. What percentage of students scored higher and how many scored lower?

Solutions to Chapter 1 Review Questions

1. See the table in the answer for question 2.
2. **Answer is c)** Supplemental oxygen should be given to COPD patients with a **goal of SaO₂ of > 90% and PaO₂ of > 50 mmHg**. This has shown to decrease mortality and morbidity. Additional parameters include **PaO₂ of 55-59 mmHg if Cor pulmonale, polycythemia, or shortness of breath during exertion are present**. This is makes it a **tertiary form of prevention** due to the aim of **decreasing mortality and morbidity**. The table below explains the three types of disease/condition prevention. See page 1 for more details.

Type of prevention	Description	Case examples
Primary	Prevent disease from developing	Smoking **prevention** campaigns
Secondary	Early detection of an existing disease with the aim of reducing complications and severity	**Screening** at regular physicians visits
Tertiary	Decrease the effect of the disease	**Treatment** with Supplemental O₂

3. **Answer a)** Remember the **Triad of Disease transmission**. This includes the agent, host, and environment. In this case the environment is the prison which is known to have more TB, the agent. See page 2 for more details.
4. An **endemic** disease/condition is a condition that is present, in a stable, oscillating, or decreasing amount in a population. A **epidemic** is a disease/condition that surges through a population in a certain area. **Pandemic** are epidemics on a more global scale. See page 4 for more details.
5. **Phase II** clinical trials involve **smaller** numbers of people than phase III. Phase II aims to check the efficacy and safety profile in affected patients using just the new treatment, while **Phase III** checks the efficacy and safety profile of the **new drug versus**

controls such as the current treatment. See page 15 for all information on clinical trials.

6. The null hypothesis is the hypothesis of no difference. When doing a study your goal is to reject or fail to reject the null hypothesis. For example, if you want to see if smoking causes lung cancer then your null hypothesis would be that smoking does not cause lung cancer. If your data shows there is a link (which in this example it obviously would) then the null hypothesis is rejected, and the alternate hypothesis would fail to be rejected. This alternate hypothesis in this case would be smoking may be linked to lung cancer. Remember, in statistics **you cannot prove anything, only fail to disprove**.

7. **Answer b)** This a common question I get from people who took the their exam for the MCCEE. The same concept applies to the USMLE.

$$Z = \frac{x - \bar{x}}{\sigma}$$
$$Z = \frac{353 - 271}{50}$$
$$Z = 1.64$$

With a Z-score of 1.64 you just need to look up in a table in the **appendix.** You can see in that table the falls within the 95th percentile, choice b). In terms of trivia, according to 2014 CaRMS data the average unmatched score is 328, so this person has a chance of matching in CaRMS assuming the rest of the application is strong. See page 12 for more details

8. There are two parts to the question:

    a.

In the first you are asked to calculate the margin of error (ME) which can found on page 11.

$$ME = Z * SEM$$

Remember that an approximation for a CI of 95% is 1.96.

$$ME = 1.96 * SEM$$

$$ME = 1.96 * \frac{\sigma}{\sqrt{n}}$$

In this case the SEM is known as 3.

$$ME = 1.96 * 3$$

$$ME = 6$$

Your confidence interval becomes: CI = 59 +/- 18

b. In the second part you need to recognize that the Z-score is (99-59)/30 which is 2. This is the 97th percentile as shown below. On the exam you will likely not be given a Z-score, but **will need to know the percentages in the areas coloured in the diagram below**.

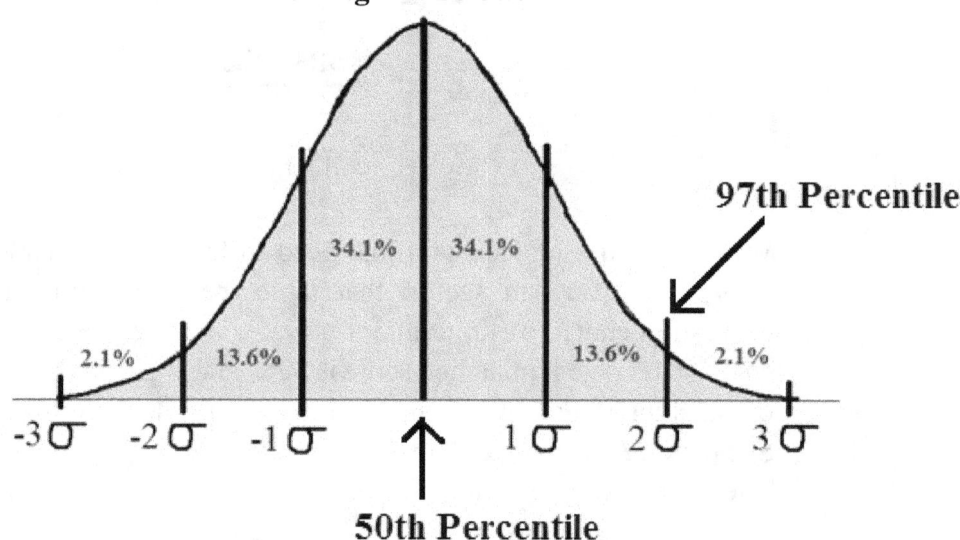

9. Recognition of skew is important for your exams.

a. **Answer is IV)** In this case the skew **tail is off to the left meaning negative distribution or choice IV**. See the diagram below. Remember that **Gaussian (standard) distribution** is **shaped like a bell** with the median = mode = mean.

b.   **Answer is II)** An easy way to remember this is that the **mean is most affected by extremes and will be near the tail**. The mode will always be on the other extreme. The median is in the middle. The diagram below will illustrate the point. Keep in mind that numbers on a line such as this always get larger when moving to the right. Look at page 6 for more information.

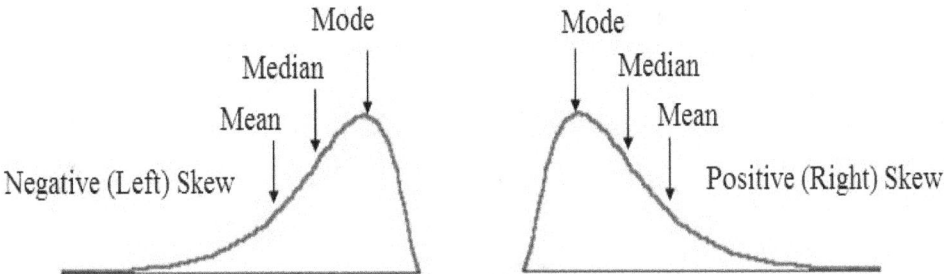

10. **Answer e)** Remember that the **median** is the **middle value in a set**, but the value does not have to be in the set. In this case there are an even number of values and the two closest to the middle are 76% and 78%. You just take the average in between them which is **77% = median.** The mean is the sum of 97%, 32%, 76%, 76%, 88%, and 79% which is 283, divided by the number of elements (4) which is **71% = mean.** This is an example of how the median was much less affected by the extreme value of 32% as the median > mean, the same goes for the **mode = 76%.** Check page 5 for more details. **This question may seem simple, but it is a possibility for your exams.**

11. **Answer c)** To answer this question you will need to know that **accuracy = validity** and **precision = reliability**. In this case the new test is correct 97% of the time compared to 93%. This is more accurate. The standard deviation is larger with the new test, making it less precise. See page 7 for more details.

12. The key to this question is that the Z-score is -1 = (75-100)/25. This means it is 1 SD below the mean. This is how the two concepts correlate. At **-1 SD there are 84.1 percent** of people who have a higher score, while **15.1 % have a lower score**. See the figure below for a graphical representation.

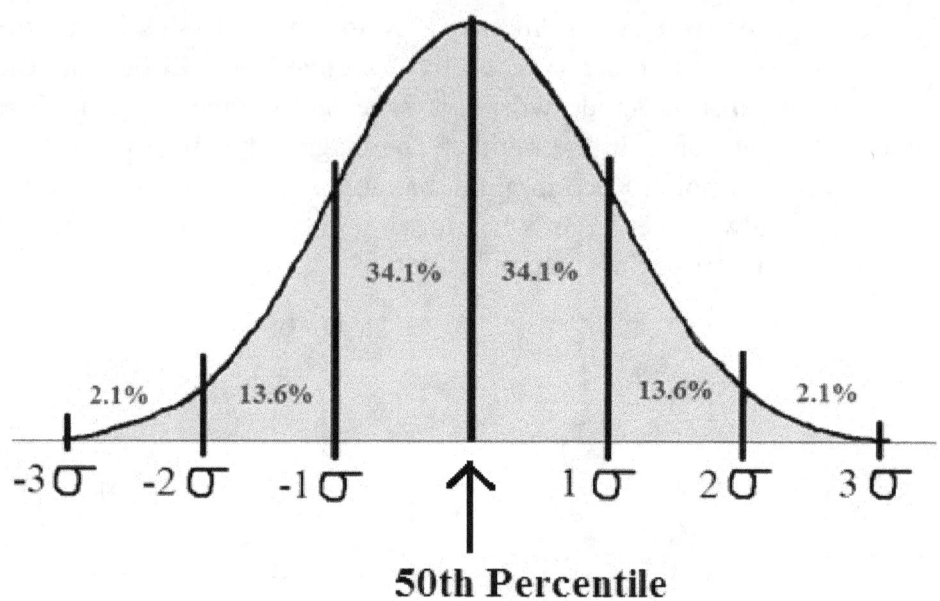

Chapter 2: Measurement Elements

Prevalence, Incidence, and Attack Rate

The prevalence of a disease is different from the incidence rate and attack rate. The definitions are simple enough, but are easy to confuse on exams.

The _____ _____ typically measures *new* cases of disease of interest in a set period of ____.

incidence rate, time

The attack rate is different in that it does not have a specific ____ frame associated with it. __ _____ people who become stricken with the ailment of study.

time, at risk

The prevalence measures the number of individuals with a disease in a population. It says _____ about the number of people who get the disease each year. Both incidence and prevalence are typically measured per ____ people.

nothing, 1000

Note: It is important to recognize that although these are straightforward definitions, try to think of how these concepts will be used on an exam. Generally you can rule out possible answer choices by just reminding yourself of these basic definitions.

Note: There is a method called sensitivity analysis which can look at experiments and trials under certain conditions (such as what if the attack rate is 10% higher or lower). This can make data more robust as long as other statistical parameters remain significant.

Standard Table

	Condition (ie. disease) truly present	Condition truly absent	Totals
New Test positive	True Positives (TP)	False Positives (FP)	New test positive results total
New Test negative	False Negatives (FN)	True Negatives (TN)	New test negative results total
Totals	Total number of people with disease	Total number of people without disease	Total population studied

This is a standard table you will need to know how to construct. It will be useful in many different calculation scenarios.

The table above is something you need to construct with the ____ _____ test or knowledge of the disease being present across the top. The new test is placed in the ___ ____ column.

gold standard, far left

Most of the findings in the chart will be intuitive. For instance a ____ _____ occurs when the gold standard test and ___ ____ are both positive. A false positive occurs when only the new test detects disease.

True Positive, test

The sensitivity of a test is good at ruling a condition ___, but has now bearing on ruling it __. Formulas will be on the next page

out, in

The _____ of a test and aims to rule a condition in. Tests can be _____, but not specific and vice versa. Formulas for both are on the next page.

specificity, sensitive

Sensitivity, Specificity, NPV, PPV

$$\text{Sensitivity} = \frac{TP}{TP + FN}$$

$$NPV = \frac{TN}{FN + TN}$$

$$\text{Specificity} = \frac{TN}{TN + FP}$$

$$PPV = \frac{TP}{FP + TP}$$

$$\text{Positive Likelihood Ratio} = \frac{\text{Sensitivity}}{1 - \text{Specificity}}$$
$$(LR^+)$$

$$\text{Negative Likelihood Ratio} = \frac{1 - \text{Sensitivity}}{\text{Specificity}}$$
$$(LR^-)$$

Sensitivity, specificity, negative predictive value (NPV), and positive predictive value (PPV) are all common computational questions, specifically for the USMLE exams; however, they are important to understand in general.

The NPV of a test and PPV of a test are arguably _____ *directly* useful in determining if a disease is present or not (see the end of chapter questions). The NPV is the chance of a condition being _____ if the test is negative. The ___ is the chance of the disease being present if the test is positive.

more, absent, PPV

There is one catch with the NPV and PPV. They are both _____ on a particular population of study. If you go to a different population of people, your NPV and PPV _____ ___ apply to that area.

dependent, will not

Note: If the equations are confusing, take another look at the table on the previous page. If they are still confusing, there will be practice questions at the end of the chapter.

Liklihood Ratios

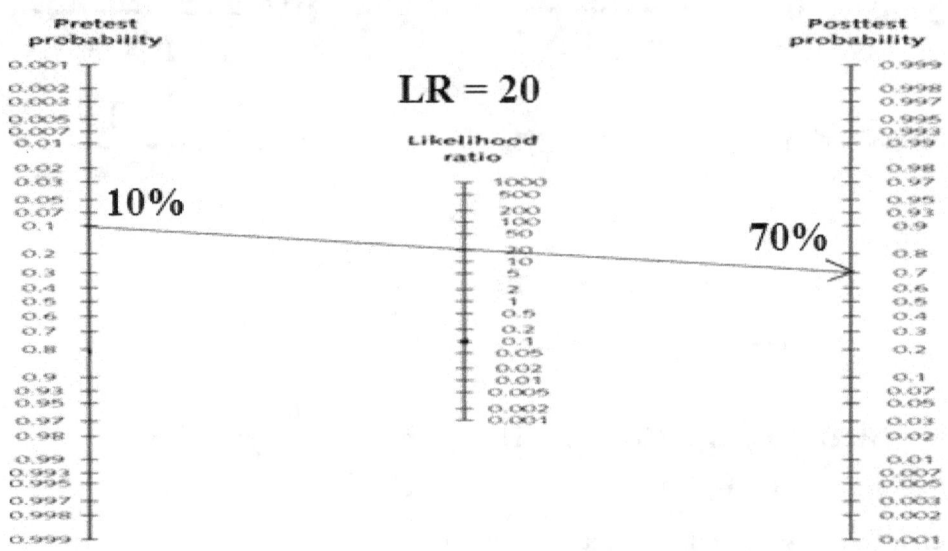

To apply the likelihood ratio to a population you need a nomogram such as above.

A positive likelihood ratio (LR$^+$) is akin to _____ in that it is the likelihood of a disease being present if the test is _____. The difference is the LR$^+$ can be applied to any population assuming the pre-test probability is known. The negative likelihood ratio (LR$^-$) is similar to a NPV. Both relate to he probability of a disease being present when a test is _____, but the LR$^-$ also can be used with any population.

PPV, positive, negative

In the example above, the _____ probability for a disease is 20%. For example, 20% of young adults with acute onset _____ pain will have appendicitis. If you have a positive test that has a LR$^+$ of 20, then you draw a line through 20% pretest _____ and LR 20. You then see what the post-test probability is. In this case it is 70%, meaning that there is a 70% chance that the condition in question is present.

pre-test, abdominal, probability

P value

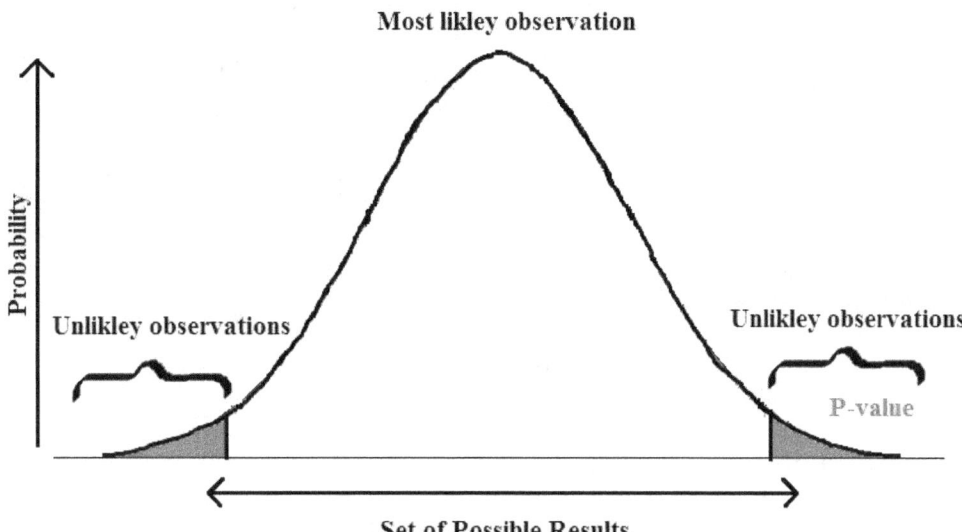

Mathematics is a pursuit that studies patterns and makes arguments to prove those patterns. The P value is a necessary, but not sufficient, part of the argument.

The P value is the likelihood that the observed study findings are more _____ than those found in the study assuming the null hypothesis is ____. You can see in the graphic above how the P value denotes the extremes of the curve. It says _____ about the likelihood of Type I error.

extreme, true, nothing

Note: A type I error will be discussed later, but it is the error of rejecting the null hypothesis when it is actually true.

The P value has an *arbitrary* cut off to help tell if the ____ hypothesis can be rejected. This is typically less than ____. Anything above generally suggests the null hypothesis should be _____ as it suggests your data is not a true representation due to random sampling error. For example if you obtain a P value of 0.02, you would expect a 2% chance that you would obtain the findings, or more extreme findings, in your study due to random sampling error if H_0 is true.

null, 0.05 (5%), accepted

Statistical Power

Actual truth

Study results		H_1 True	H_0 True
	H_1 True	**Correct**	False Positive or Type I Error (probability = α)
	H_0 True	False negative or Type II error (probability = β)	**Correct => Power (Probability = 1- β)**

Unlike the P value, the power of a sample directly relates to the probability of making a Type II error.

The _____ of a study is the probability that you have correctly decided that the null hypothesis should be rejected. The formula is shown in the table above. The power of a study can be increased by increasing the _____ _____. The lower the power the less likely you are to observe small benefit or harm from a medication or procedure.

power, sample size

Most values of power are *arbitrarily* set at _____. Alpha (α) is generally set at 0.05 or 0.01 (95% or 99%) meaning changing the sample size does not affect Type I errors. Researchers must also choose if the test is one or two _____. If it is two sided you need a _____ sample size, but it can tell you if a new treatment is harmful as well as beneficial. One sided tests are more common due to less _____. Power also reduces _____ _____ which is when a test with a positive result is exaggerated because not enough data (not enough power) was collected. For example a study of a drug that treats diabetic nephropathy at 50% power might say that the patient will regain 90% of their kidney function, but when adequately powered the results may only be 10%. Still a positive result, but drastically different.

80-90%, sided, larger, cost, truth inflation

Number Needed to Treat and Harm

	Condition (ie. disease) truly present	Condition truly absent
Exposed	True Positives (TP) = **a**	False Positives (FP) = **b**
Not Exposed	False Negatives (FN) = **c**	True Negatives (TN) = **d**

$$AR = \frac{c}{c+d} - \frac{a}{a+d}$$

$$NNT = \frac{1}{AR}$$

$$NNH = \frac{1}{ARI}$$

AR = attributable risk, NNT = number needed to treat, NNH = number needed to harm, ARI = attributable risk increase

_____ ____ reduction is the risk associated with exposure to the risk factor minus those who are not exposed. For instance people who ____ have a higher chance of lung cancer than those who do not; however, you can get lung cancer if you do not smoke. This tells you how many people you would expect to get ____ ____ in the smoking group even if they did not smoke.

Absolute risk, smoke, lung cancer

The number needed to treat (NNT) is the number of people who need to receive _____ in order for one person to _____ from the treatment. The number needed to harm (NNH) is the same, but measures ____ _____. The absolute risk increase (ARI) is found the same way as attributable risk reduction except the table will measure a particular adverse effect.

treatment, benefit, adverse events

____ _____ _____is the possible benefit minus the possible harms. An example would be chemotherapy shrinking a tumor, but causes organ toxicity.

Net clinical benefit

Note: for exams treat attributable risk (AR) and absolute risk reduction (ARR) are the same

Relative Risk and Odds Ratio

	Condition (ie. disease) truly present	Condition truly absent
Exposed	a	b
Not Exposed	c	d

$$RR = \frac{a}{a+b} \cdot \frac{c+d}{c}$$

$$OR = \frac{a \cdot d}{c \cdot b}$$

RR or OR < 1 => negative association

RR or OR > 1 => positive association

RR = relative risk, OR = odds ratio; these tests are weaker than a randomized controlled trials and are considered observational studies, they do not control for the independent variable and are not meant to influence practice changes – they provide a basis for a future RCT. Very important!

Relative risk (RR) is the probability of an event occurring in group _____ to substance that may cause disease compared to a non-exposed group. It is calculated in cohort studies (discussed in Chapter 3). If the RR is ___ then that means there is no association. When RR is _____ than one it implies an association between exposure and disease.

exposed, one, larger

The ____ ____ (OR) is calculated from a case-control study (discussed in Chapter 3). It is the odds of finding someone with an exposure compared to no exposure when both groups have the _____ disease. The OR can approximate the RR if the disease is rare.

odds ratio, same

The hazard ratio (HR) is a value that it relates the number of adverse events in a _____ group compared to the number of adverse events in the control group. For instance a HR of 2 indicates that ____ as many adverse events are occurring in the treatment group, while __ would indicate only half of these events are occurring in the treatment arm. Hazard ratios differ from RR since they can be taken at any time. This can be used to the advantage of a pharmaceutical company by stopping data gathering when the time is optimum. Another reason why it is important to not stop a trial early.

treatment, twice, 0.5

Chi Squared, T-tests, and ANOVA

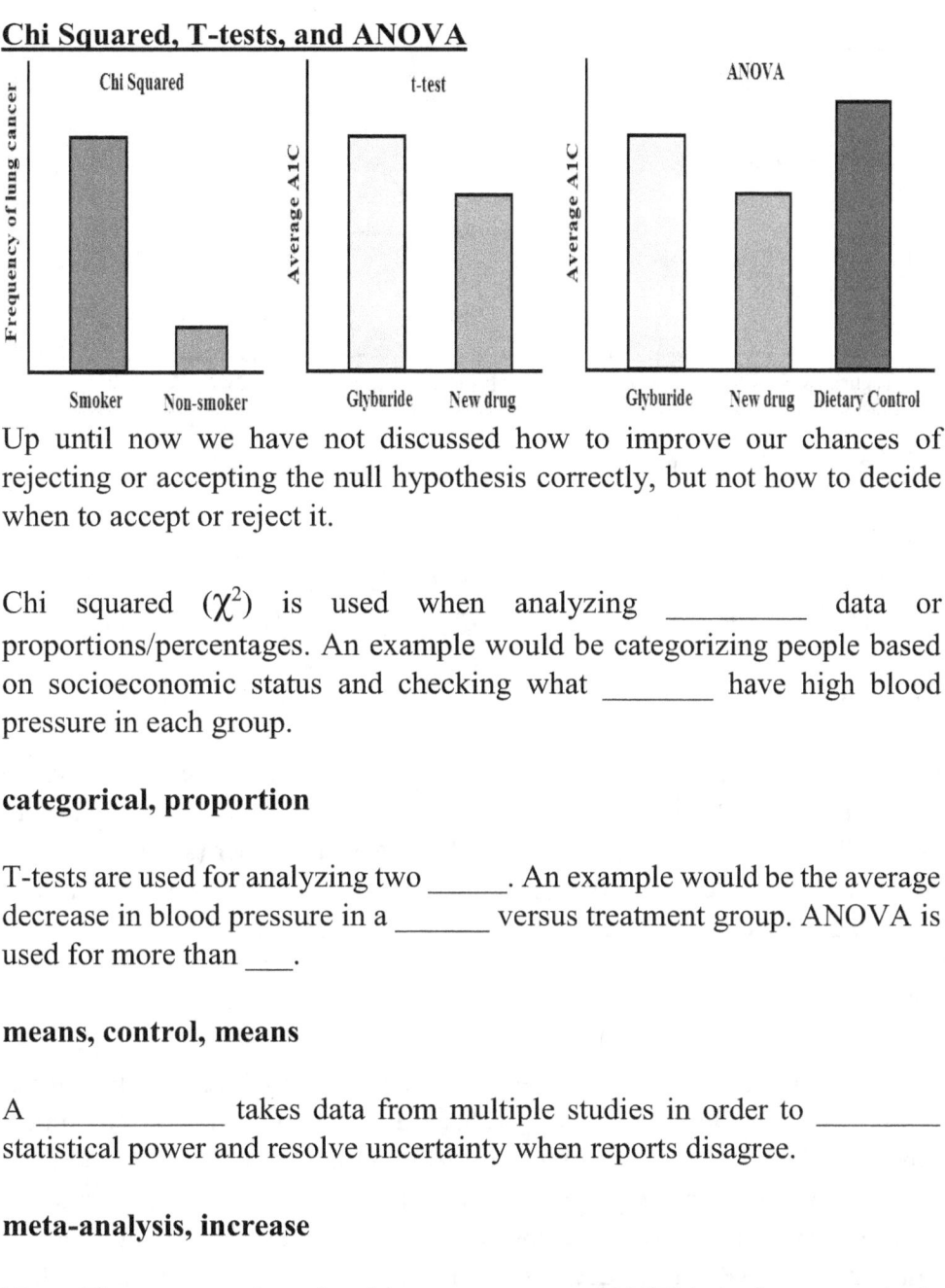

Up until now we have not discussed how to improve our chances of rejecting or accepting the null hypothesis correctly, but not how to decide when to accept or reject it.

Chi squared (χ^2) is used when analyzing _____ data or proportions/percentages. An example would be categorizing people based on socioeconomic status and checking what _____ have high blood pressure in each group.

categorical, proportion

T-tests are used for analyzing two _____. An example would be the average decrease in blood pressure in a _____ versus treatment group. ANOVA is used for more than ___.

means, control, means

A _____ takes data from multiple studies in order to _____ statistical power and resolve uncertainty when reports disagree.

meta-analysis, increase

Note: The computations for chi square, t-tests, ANOVA, and meta-analysis are too complicated for standardized exams; however, you do need to know when they are used.

Pearson Correlation Coefficient

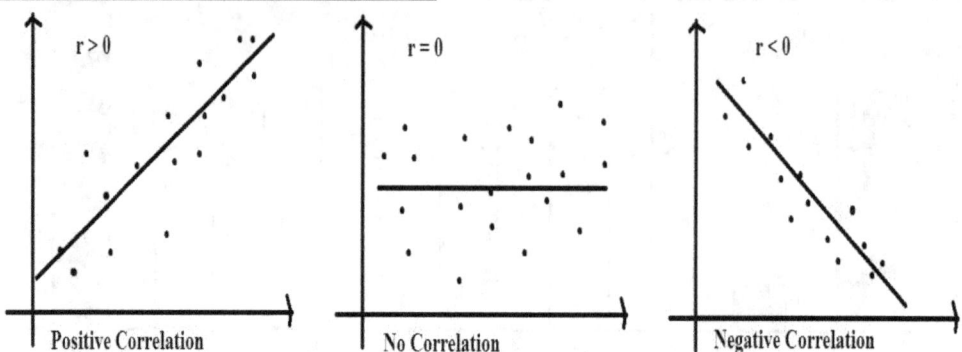

On USMLE exams sometimes you will be shown a picture such as above and be asked about the relation.

The Pearson correlation coefficient is a measure of the _____ correlation between two variables. In other words it tells you how strongly or weakly ____ variables (such as smoking and_____ _____) relate to one another.

linear, two, lung cancer

The _____ relation is 1 or -1. An _____ relation shows that as one value increases the condition in questions does as well. A negative relation instead shows that the _____ of one factor will decrease the other (such as amount of cigarettes smoked and life expectancy). There are other types of relations other than linear, but they are not as important for the exams.

perfect, increase

Note: Surrogate markers are objective values that improve that the patient does not notice and are a marker of disease, but do not improve outcomes. An example would be changing the A1c of a diabetic from 8.0% to 7.0%. This would not change patient oriented outcomes. Beware of surrogate markers! If a paper does not compare these markers to patient oriented outcomes (less sick days, decreased hospitalizations) then that is a red flag the study is only looking at a surrogate markers!

Receiver Operating Characteristic Curve (ROC)

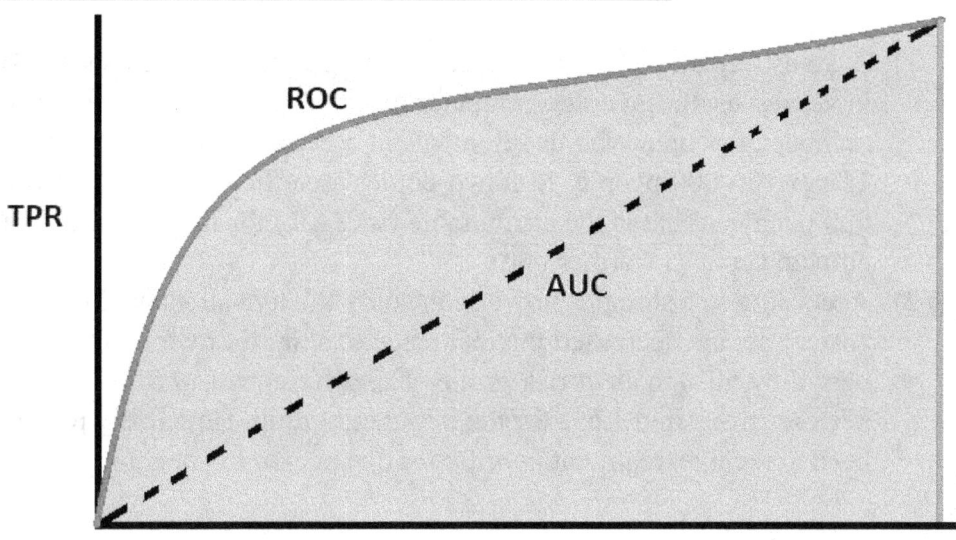

TPR = true positive rate, FPR = false positive rate, AUC = area under curve (denoted in gray), ROC is in red

The receiver operating characteristic (ROC) is a complicated method that has only one primary aspect you need to know: the _____ the size of the AUC the better the test is. It depicts the balance between sensitivity and 1-specificity. The AUC gets smaller the closer the curve gets to the dotted line or the more linear it becomes. On exams you will likely be asked which test is the most robust and pick between many curves, just pick the curve that is the _____ up and to the left of the dotted line.

greater, furthest

_____ _____ is another complex method of analysis, but it is used in observation studies exclusively. It attempts to _____ a randomized control trial (the gold standard for a study that CAN lead to practice changing results unlike observational studies). It accounts for covariates and through complex mathematics beyond the scope of this book. Another measuring technique is the standardized mortality ratio (SMR) which is a _____ of deaths in study population to deaths in the general population to help look at causality.

Propensity scoring, mimic, ratio

End of Chapter 2 Review Questions

1) A study finds that the risk of developing multiple sclerosis (MS) increases as the average temperature in a region decreases. The number of people who develop MS in a certain warmer climate is 15 per 100 000 people. In much colder area the rate is 30 per 100 000 people. What is the attributable risk (AR) for living in a colder climate versus a warmer one?

2) According to a shingles prevention study the live attenuated Herpes zoster vaccine decreased the incidence of shingles over a three year period by 50% in an at risk group. The difference was 3.3% for the placebo group and 1.6% for the treatment group. How many people need to receive treatment in order for one person to not get shingles?
   a. 1
   b. 43
   c. 2
   d. 59
   e. 14

3) A new study is performed to see if eating dark chocolate, regardless of amount, would increase life expectancy. The study follows 11 individuals in the dark chocolate group and 17 in the control group. The study follows what proportion of people reach 75 years old. The P value is 0.001. What kind of analysis should be used to decide if the null hypothesis should be rejected or fail to be rejected?
   a. T-test
   b. Chi squared
   c. Meta-analysis
   d. ANOVA

4) Taking a look at the study in question 3) what would you think the power of the study is? What is the significance of this?

5) What does the P value in question 3) mean?
   a. There is a 0.1% chance a Type I Error was made
   b. The null hypothesis should be rejected
   c. The null hypothesis should be accepted
   d. There is a 0.1% chance a Type II error was made
   e. There is 0.1% chance that the actual results are more extreme

6) A new disease has evolved to being symbiotic with mosquitoes and is transmitted by them. It causes individuals to suffer from a severe

anemia and lose all feeling in their noses. Researchers have dubbed it "numbing nose." Companies are attempting to work on both a method of detection and a cure. One company creates two equal groups, one group where those are known to have the disease and one control with a total of 1000 people. They obtain a sensitivity of 86% and a PPV of 93% (Hint: Make a table).

    a. How many false negatives are there?

    b. What is the positive likelihood ratio?

7) A study is done to see how life expectancy related to the amount of time spent in front of a television. The following scatterplot was obtained.

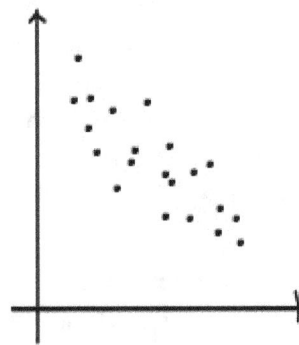

    a. What kind of relationship is it?

        i. Parabolic

        ii. Linear

        iii. Exponential

        iv. Sigmoidal

        v. Hyperbolic

    b. What is the likely correlation factor?

        i. $r = 0.9$

        ii. $r = 0.01$

        iii. $r = -0.8$

        iv. $r = -0.3$

8) A 63 year old man comes into the clinic for a routine examination. He has diabetes mellitus and hypercholesterolemia, but both are well controlled. His BP today is 143/87. He asks about getting a PSA screen because one of his friends has it. He denies any urinary signs/symptoms. You tell him that 1/7 men will get prostate cancer in their lifetime. If he actually has prostate cancer then there is a 26% chance that he will test positive. If he does not have prostate

cancer he has a 10% chance of testing positive. What is the probably of this patient having prostate cancer if his test comes back positive?
   a. 30%
   b. 17%
   c. 60%
   d. 92%
   e. 1%
   f. 43%
   g. 6%

9) What is the most important consideration in a screening test?

10) A group of scientists is working on a Case-Control study of Liddle's syndrome. The syndrome is very rare and causes high blood pressure, metabolic alkalosis, and hypokalemia. Which of the following regarding the odds ratio (OR) is true?
   a. The OR id not an approximation of relative risk
   b. The relative risk is generally much easier to obtain
   c. The incidence of the disease must be known to calculate and odds ratio
   d. The OR approximates the RR when the disease is rare

11) A prospective cohort study is done calculate the relative risk of radon exposure with lung cancer. A group of 1000 people were followed and 100 developed lung cancer with 70 of those being exposed to radon. There were 300 people exposed to radon who did not become diseased. What is the relative risk associated with radon exposure?

12) A large study is performed in the general population over a period of several years to examine Alzheimer's disease. From the study a group of medical students conclude that there is an age related increase in the risk of Alzheimer's disease. The data is below.

Age Group	People with Alzheimer's (%)
40-49	1
50-59	3
60-69	8
> 70	15

What do you say to this?

13) The pretest probability of someone having HIV infection is 0.3% and the sensitivity and specificity of ELISA testing are 99.1% and 99.7% respectively. If a test is positive, what is the post-test probability (Hint: Use appendix B)?

Solutions to Chapter 2 Review Questions

1)

Remember that **attributable risk (AR)** is the difference in incidence between the at risk population and the non-exposed population. This allows the finding of how many people who became afflicted with the disease would have become diseased even without the exposure. In this case there are $\frac{15}{100000}$ becoming afflicted MS each year in southern regions while $\frac{30}{100000}$ are becoming afflicted yearly in northern regions. The $AR = \frac{30-15}{100000} = 0.00015$ or 0.015%. This means the AR is 0.015%. Check page 29 for more information.

2)

Answer d) The key to this question is understanding that it is looking for the **NNT (number needed to treat)**. This is found by the equation: $NNT = \frac{1}{ARR}$. The **ARR (absolute risk reduction)** is 3.3%-1.6% (control − treatment) which is 1.7% (0.017). You plug this into the equation $NNT = \frac{1}{0.017} = 59$. Thus 59 people will need to be treated before one person sees a benefit. This is a good point on how a large decrease (50%) in incidence still means that a **large number of people will need treatment when a disease is not incredibly common.** Check page 29 for more information.

3) **Answer b)** When deciding what test to use remember to see what the study is using to measure differences. In this case **proportions** are being used to measure **categorical data**. This means that **Chi squared** should be used. ANOVA and t-tests should be used for means (ANOVA for > 2 means). Meta-analysis is used when multiple studies are used. Check page 31 for more details.

4) The **power** of a study is **directly related to the sample size.** That being said the larger the sample the great the statistical power. In this question only a couple of dozen people were studied. The **power is likely so low that a Type I error is a real possibility. Check page 28 for more details.**

5) **Answer e)** Remember from page 27 that the **P value is the likelihood that the observations are true if the null hypothesis is**

true. You cannot make any decisions about the null hypothesis or predict if you made a type II error, which is related **to power** and **sample size**.

6) This question has two parts but it is easy to visualize if you make a table.

a.

In this case there are two equal groups from a total of 1000, so 500 each. If this parameter is not given you can assume two equal groups and pick an easy number such as 1000 to work with. To get the specificity we need to know the TP. Remember that the FN can be calculated from the TN if we know how many there are. We know that $Sensitivty\ (SE) = \frac{TP}{TP+TN}$. TP +TN (look at the table below) is 500. SE is 86% so: $TP = 86\% * 500$, which is 430.

	Condition (ie. disease) truly present	Condition truly absent	Total
Test Postitive	True Positives (TP) = 430	False Positives (FP)	
Test Negative	False Negatives (FN) = 70	True Negatives (TN)	
Total	500	500	1000

Knowing this we can look at the PPV. $PPV = \frac{TP}{TP+FP}$. We do not know how many positive tests there are, so TP+FP is unknown. This means we need to do some algebra (fun I know).

Rearranging: $PPV(TP + FP) = TP$ then becomes

$$FP = \frac{TP(1 - PPV)}{PPV}$$
$$FP = \frac{430(1 - 0.93)}{0.93}$$
$$FP = 32$$

Thus there are 32 false positive results.

	Condition (ie. disease) truly present	Condition truly absent	Total
Test Positive	True Positives (TP) = 430	False Positives (FP) = 32	
Test Negative	False Negatives (FN) = 70	True Negatives (TN) = 468	
Total	500	500	1000

b.

The likelihood ratio is more straight forward at this step, but we do need the specificity. $Specificity\ (SP) = \frac{TN}{TN+FP}$

$$SP = \frac{468}{500}$$
$$SP = 93.8\%$$

The positive LR is $LR^+ = \frac{SE}{1-SP}$

$$LR^+ = \frac{0.86}{0.062}$$
$$LR^+ = 13.4$$

See pages 24-25 for more details.

7) This is another two part question. Check page 32 for details.

    a. **Answer II)** I know some of you are furious right now and wondering where in the book I covered anything other than a linear shape. Keep in mind, you will asked questions on exams you have not studied. Just stay calm. In this case a linear relation works well. Just in a question does come up, I will add what non-linear shapes look like.

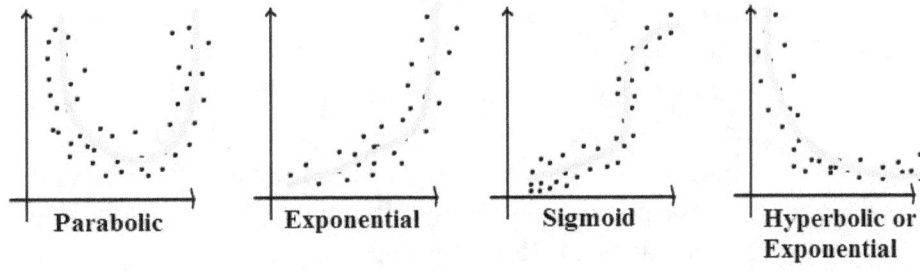

b. **Answer III)** Remember that a **negative correlation** is one with a **negative slope**, so we are looking for a negative R value. This one is fairly steep so close to -1 should work.

8) **Answer a)** The key here is to realize that you are looking for the **PPV.** You are told **through description** that the **sensitivity is 26%** and the **specificity is 10%.** Constructing a standard table can help to visualize the problem, but it can be solved without. We know that there 1/7 men will get prostate cancer. This means that in a population of 1000, 143 will have the disease and 857 will not. The number of TP from this can be seen in the Sensitivity of 26%. TP = 143 * 0.26 or 37. The number of false positives can be found by FP = 857 * 0.10 or 86. Plug this into the equation:

$$PPV = \frac{TP}{TP + FP}$$
$$PPV = \frac{37}{37 + 86}$$
$$PPV = 30\%$$

	Condition (ie. disease) truly present	Condition truly absent	Total
Test Postitive	True Positives (TP) = 37	False Positives (FP) = 86	
Test Negative	False Negatives (FN) = 106	True Negatives (TN) = 771	
Total	143	857	1000

As you can see, the PPV is quite low and means this is not a great method of screening. The other issues with prostate screening is the low number of people who benefit from early treatment and the inability to differentiate between indolent cancers, aggressive, and BPH (benign prostatic neoplasia). Check pages 24-25 for more details.

9) Just look at your answer from the previous question and you can see that PPV is the most important aspect behind a screening test. Logically you can tell that if a screening test has a PPV of 30%. That means a positive test does will be a true positive only 1/3 of the time. Check pages 24-25 for more details.

10) **Answer d) The RR is a calculation that can only be derived from a cohort study;** however, **if the disease is rare the OR can closely**

approximate it. **Cohort studies are difficult and expensive,** so case control studies can be performed instead. These are discussed in the next chapter. The **incidence of a disease must be known for a RR, but is not required for the OR**. See page 30 for more information.

11)

The equation for **relative risk (RR)** is $RR = \frac{a}{a+b} \cdot \frac{c+d}{c}$. In the table below you can see what each letter represents. You are given a few numbers already, a+c=100 and a = 70. We also know that c=300. This gives us: $RR = \frac{70}{370} \cdot \frac{630}{30} = 3.97$. That means that roughly four times as many people who are exposed to radon will get lung cancer compared to no exposure in this fictions study.

	Condition (ie. disease) truly present	Condition truly absent	Total
Exposed	a = 70	b = 300	a+b = 370
Not Exposed	c = 30	d = 600	c+d = 630
Total	100	900	1000

12) Remember that **when prevalence is used you cannot tell what the incidence is**. For that statement to be correct you would need to know how many people are diagnosed with Alzheimer's each year in each age group.

13) First you will need the **likelihood ratio** as shown here:

$$LR^+ = \frac{SE}{1 - SP}$$
$$LR^+ = \frac{0.991}{1 - 0.997}$$
$$LR^+ = 330$$

Using the **nomogram in Appendix B we see the post-test probability is 90%.** This makes the disease highly probable, but a different test (ie. Western blot) should be used to confirm since there is still a 10% chance of not being correct. This is especially true with diseases that have harsh treatments and consequences such as HIV.

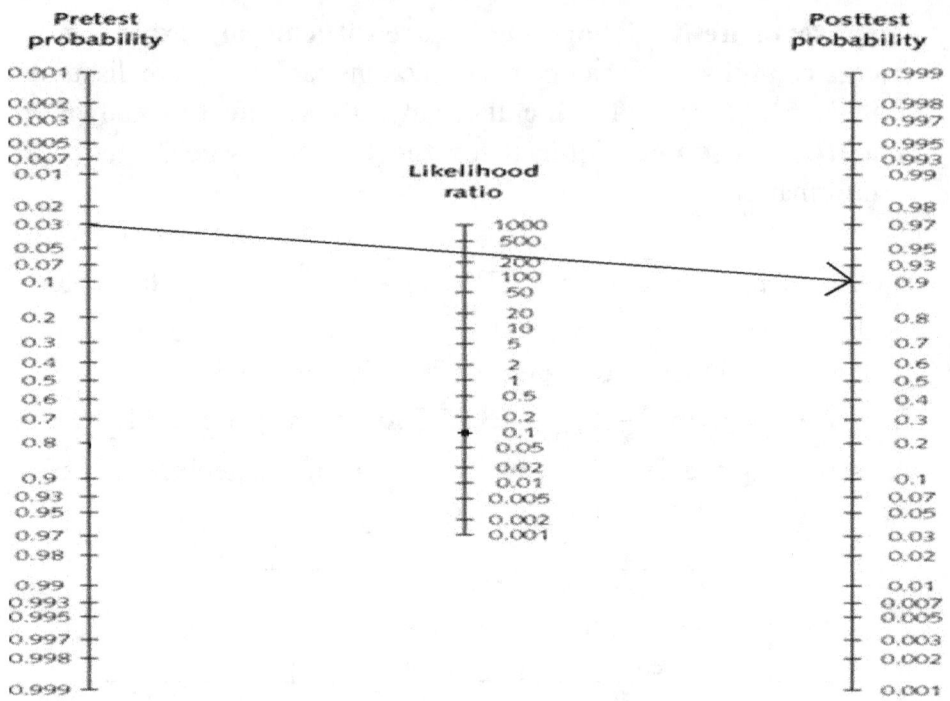

Chapter 3: Important Study Types, Bias, and Error
Crossover and Crossectional Studies

Group 1: O ⟶ X ⟶ O

Group 2: X ⟶ O ⟶ X

Each X represents a treatment group while each O represents a control group.

A _____ study allows for each patient to receive both the primary treatment and the _____ treatment. This study has the advantage of eliminating _____ variables (discussed later) and require fewer numbers of people than non-crossover designs.

crossover, control, confounding

A weakness of crossover studies is the _____ effect. This means that the effect of the first drug could carry over during the treatment of the next treatment. _____ periods can help prevent this, but are not always possible.

carryover, Washout

Cross-sectional studies are studies taken at a _____ point in time. They help to reveal the _____ in a population. Data on confounding _____ cannot be obtained in this way, but these are _____ to perform.

specific, prevalence, variables, inexpensive

Cohort and Case Control Studies

The way this author remembered that case-control and odds ratio were linked was thinking of a dice (playing the odds) falling out of a briefcase.

A Cohort study are the only studies to measure _____ _____; however, they are more _____ and time consuming to perform. Cohort studies also measure the disease _____. Prospective studies look at a risk factor exposure and you wait to see who becomes diseased. _____ requires the review of the past records then looking at the risk factors and who becomes diseased.

relative risk, expensive, incidence, retrospective

By contrast _____ _____ studies measure the odds ratio. Instead of looking at exposures and then see who becomes diseased, you look at who has the disease and what the _____ are. This is less accurate than cohort studies, unless the disease is very _____.

case control, exposures, rare

A ____ _____ studies look at identical twins and assess for the risk of a condition developing. Adoption studies are the same but look more at _____ children compared to unadopted.

twin concordance, adopted

Type I and Type II Error

Actual truth

Study results		H_1 True	H_0 True
	H_1 True	Correct	**False Positive or Type I Error (probability = β)**
	H_0 True	**False negative or Type II error (probability = α)**	Correct => Power (Probability = 1- β)

Type I and type II errors have nothing to do with the P value. Remember that for your exams.

A type I error is the probability of _____ the null hypothesis when it is _____, otherwise known as a false positive. Remember the power of the study is generally arbitrarily set at ___, meaning the probability of a Type I error should be less than 20% as long as the study is adequately _____.

rejecting, true, 80%, powered

A ____ __ _____ is the probability of accepting the null hypothesis when it is ____. This is a false negative whose probability is denoted ____.

type II error, false, alpha

Note: I cannot stress this enough, but a P value from a SINGLE study WILL NOT give you any indication of the probability of a type I or II error.

Confounding Variables and Effect Modification

OR/RR from crude data

OR/RR Strata 1 ≠ OR/RR Strata 2
Effect Modifier (Truth)

stratified analyisis

OR/RR crude ≠ OR/RR corrected
Confounding (error of study)

Strata 1 OR/RR

Strata 2 OR/RR

Corrected OR/RR

To see effect modification and confounding variables you need to check four different RR or OR (depending on the study type): crude, strata, and the corrected.

Bias is a _____ _____ in study design or conduct of a study. There are many types. A _____ _____ is a *distortion* of effects due to an extraneous factor is accidently mixed with the actual exposure. To discover this you must break the groups into _____ and correct the OR or RR. If they are different you know you have a confounding variable. Confounders are related to _____ and outcome.

systemic error, confounding variable, strata, exposure

Confounding can be _____ by proper randomization, matching, and _____ of potential cofounders; however, restriction can cause a loss of statistical _____.

avoided, restriction, power

Effect modification is a _____ in the study and not bias, so there is no way to *correct* for it like _____ variables. Effect modification is just another factor that *modifies* the exposure. The way to discover the effect is to check if the _____ RR or OR are different between different strata (ie. a strata for men and a strata for women or smokers and non-smokers). Confounding and effect modification are not _____ _____ and can occur at the same time.

truth, confounding, stratified, mutually exclusive

Selection Biases

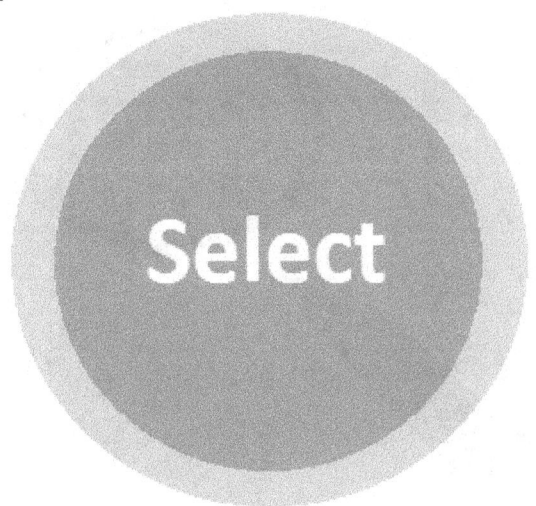

Non-representative samples of the study population (ie. poor retention).

Ascertainment, otherwise known as _____ bias is when the study population _____ from the target population. The problem is due to ___ _____ sampling. _____ bias occurs when treatment regimen depend on the severity of the patient condition.

sampling, differs, non-random, Susceptibility

_____ biases can occur if too many people are nonresponders to survey's or questionnaires and are _____ different from people who did respond. A similar bias can be seen if patients are lost to follow up are different in those who do not. This is called _____ bias.

Non-response, significantly, attrition

When only _____ patients are used a Berkson bias may be introduced. This is because hospitalized patients may ____ from the target patients. A _____ bias can be seen in populations where exposures have a long lead time leading a study to miss diseased patients who die early or recover (ie. diabetes). This skews data due to selective survival. It is also known as a _____ bias.

hospitalized, differ, prevalence, Neyman

Observational Biases

Observational biases are inaccurate measurements or classification of disease, exposure or other variable. Also known as measurement or information biases.

A _____ bias is most common in _____ studies and relates to the fact that negative outcomes are more likely to be report certain exposures than control subjects. Similar to this is the reporting bias which is _____ exposure history due to a perceived social stigmatization (ie. reporting needle sharing).

recall, retrospective, underreporting

The surveillance bias is also known as a _____ bias. It occurs due to the risk factor itself causing an increase in _____ in exposed group compared to the unexposed group. This leads to an _____ probability of detecting the disease.

detection, monitoring, increased

When observers misclassify data due to individual differences in _____ or preconceived expectations regarding the study then an _____ bias has been made.

interpretation, observer

Miscellaneous Biases

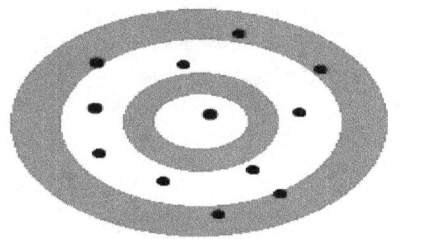

High bias

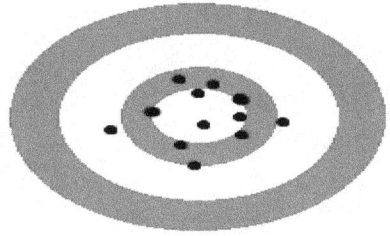

Low Bias

There are other biases that do not fall into the definitions selection or observational. These will be discussed here.

A ___-____ bias is one gathered at an inappropriate time. An example is in a study of a _____ disease where only those alive will be able to answer the questions. ____ _____ can also be confused with survival, which is known as a lead time bias.

late-look, fatal, early detection

A procedure bias can occur if different groups have are treated _____ or have a dissimilar protocol. An example can be seen with less _____ being spent on the placebo group leading to less adherence. A similar bias called the observer-expectancy effect can occur when the researchers _____ that the treatment works and this _____ the effects. Whereas the _____ _____ refers to patients who may act differently when under observation (comparable to how you drive when you see a police car nearby).

differently, attention, believe, changes, Hawthorne effect

Medicine is also well known for the _____ bias. This is the studies that get published based on results and in medicine they are inclined towards _____ results. For an intuitive example on this this works we all know about the Wright brothers, but what about all of the people who made machines that failed to produce the desired result of flight. You may think knowing the failed trials are meaningless, but why a particular method failed is important if you are to find a way to create a working one.

publication, positive

Reducing Bias

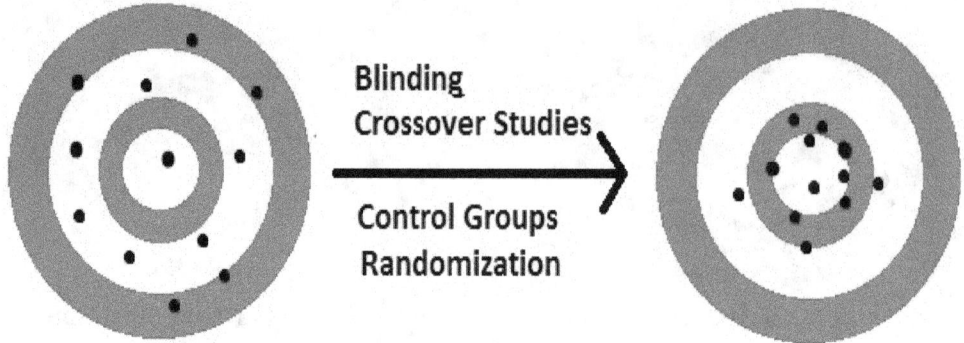

Bias can be _____ in a number of ways and many of these methods have been discussed earlier. The _____ of studies helps to limit the influence of researchers on interpretation of outcomes as does the addition of _____ groups.

reduced, blinding, control (or placebo)

As mentioned at the beginning of the chapter, _____ studies have each patient acting as their own control. This _____ confounding variables and observer bias.

crossover, reduces

By far the simplest way to reduce bias is to ensure proper _____, which reduces selection bias and _____ variables. Proper _____ also reduces confounding biases.

randomization, confounding, matching

Note: beware of ecological bias as well, which is using data from a group and applying them to the individual level

End of Chapter 3 Review Questions

1) A new study followed a group of diabetic patients over a period of 15 years. During that time the effect of A1C levels and peripheral arterial disease (PAD). They want to see if an A1C of 6.0 - 6.5%, the control group, versus less than 6.0% would lead to different outcomes. The RR from the crude data ended up being 1.04. The results are stratified into those who smoke and those who do not. In the smoking group the RR is 1.07 while in non-smokers the risk is 1.89. Using this date the corrected RR is made to 1.31. Based on this study which of the following is true?
    a. The study is affected by effect modification and confounding variables
    b. The study is affected by effect modification
    c. The study is affected by confounding variables
    d. The study is free of bias
    e. Non-smoking diabetics should have tighter diabetic control
2) Is effect modification a type of bias?
3) A study is performed to see what the effect of age is on prostate cancer. Three groups are looked at: ages 21-35, 36-50, 51-65. To achieve this urine PSA and digital rectal exams were performed on over 3000 patients. The PSA cut off here was less than 4.0 ng/mL. What type of study has been performed?
    a. Retrospective cohort study
    b. Cross-sectional
    c. Case control
    d. Prospective cohort study
    e. Crossover
4) Does the study in question three provide information on incidence, prevalence, or neither?
5) Radon is a breakdown product of Uranium and so can be found in the soil that contains this element. The radiation release from radon can lead to lung cancer. A controversy in medicine is whether radon exposure in the home can lead to lung cancer. A random group of 1000 people with and without lung cancer are chosen. Of these people 100 have lung cancer, 500 healthy people have no exposure, and 450 have been exposed to Radon in the home. What are the odds

or risk of developing lung cancer if you are exposed to Radon in the home?

    a. 0.99

    b. 1.04

    c. 1.22

    d. 2.73

    e. 1.25

    f. 0.43

6) In the previous question there is nothing to indicate how much bias affected the results. Which of the following would reduce bias in the study?

    a. Randomization

    b. Crossover study

    c. Control groups

    d. Triple blinding

    e. Changing the p value cut off to 0.01

7) With each of the following studies state which bias that is present. For convenience I will list the biases you will need here: observer, ascertainment, prevalence, procedure, Berkson, non-response, Hawthorne, lead time, attrition, recall, reporting, and surveillance

Example	Bias
i) A study is performed to see the effects of a new class of lipid lowering medication. A trial is conducted on inpatients in hospitals across the country.	a)
ii) A survey is sent out to those who have Huntington's disease. The average age of those taking the study is 41.	b)
iii) A set of surveys is sent out a population asking about antibiotic use habits such as if they finish the whole prescription as asked. 97% say yes. It is later revealed only healthcare providers took the survey.	c)
iv) A new form of chemotherapy is set for trial. The groups are randomized and blinded. The medication seemed effective, but many people in the treatment group end up dropping out due to the extreme side effects of the medication.	d)
v) A 16 year old girl does a science project on antibiotic use among households, but only asks people living in a more affluent neighbourhood.	e)
vi) To see the how intravenous drug use relates to income status several surveys were sent out to randomized strata of different income levels. It is later discovered that many people in more affluent communities did not tell the truth in fear of how others would think if they found out.	f)
vii) To discover adverse effects from a new version of acetaminophen households to see if they can remember adverse outcomes versus the benefits. People who experienced some mild abdominal discomfort were less likely than those who did not have adverse effects to report that the medication relieved their headache.	g)
viii) During a trial looking at Med A and Med B the physicians have heard that Med A causes anemia more often than Med B. The doctors make sure to run CBCs more often on people with the Med A group. They study finds Med A does cause anemia.	h)

ix) During a study researchers discover there is   i)
a significant difference between two groups. It
is found that each researcher used their own
criteria for a certain diagnosis.

x) A particular cancer has a mean survival of 5   j)
years. A new test discovers the cancer six
months earlier and with it changes the survival
to 5.5 years.

xi) A physician runs a study where he does not   k)
put in adequate controls to ensure adherence in
the placebo group.

xii) A nutritional study is performed to see how   l)
dietary habits affect mood. The physician sees
those in the study daily and asks about what they
have been eating.

8) For the following screening test what would happen if the cut off value for a normal hemoglobin concentration (to test for anemia) was to move from A to B?

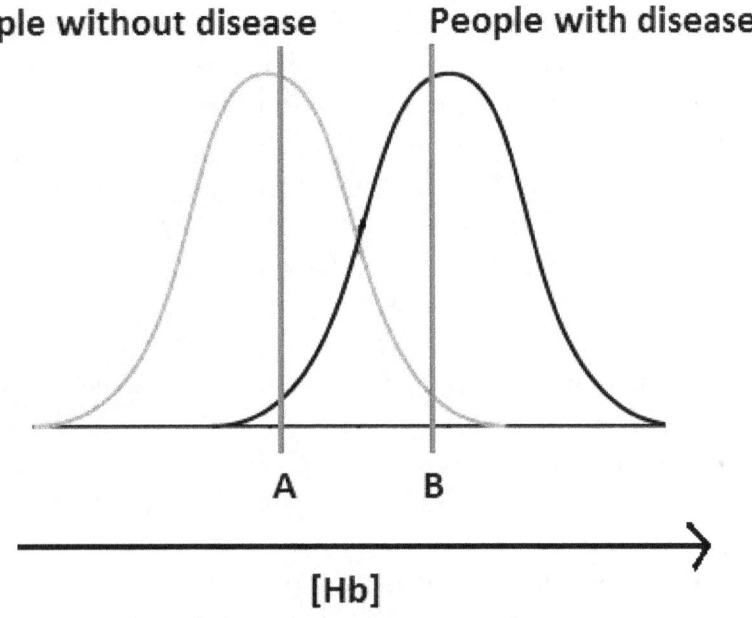

People without disease   **People with disease**

A   B

[Hb]

   a. Specificity increases, Sensitivity increases
   b. PPV increases, NPV increases
   c. Sensitivity decreases, PPV increases
   d. Sensitivity increases, specificity decreases
   e. Specificity increases, NPV increases

Solutions to Chapter 3 Review Questions

1) **Answer is a).** Remember that When the OR or RR of the strata are different from each other than **effect modification** is present. In this case the smoking group is different from the non-smoking group. **Confounding variables** are discovered when the crude OR or RR is different from the corrected RR of the stratified analysis. In this case those are also different. As you can see **confounding variables and effect modification can occur at the same time**. See page 45 for more information.

2) Remember that **effect modification** is a **truth** of the study. It is found when a **stratified analysis** reveals a difference in RR/OR between the strata. It is just a **factor** that affects the results of the study. See page 45 for more information.

3) **Answer b)** When a group of individuals are looked at during a **specific point in time** which makes this is a **cross sectional study.** For more information please see page 42-43.

4) **Cross-sectional** studies will provide information on **prevalence**, while **case-control and cohort studies** will provide information on **incidence**. See page 42 for more information.

5) **Answer e)** The key to this question is understanding this is a **case control study** which means the **odds ratio** will be used. You can see this because the question mentions it was **checking those with the disease against what exposures** they may have had. By contrast, **cohort studies** check for exposures and look at the disease outcome which uses **relative risk**. To answer this you need to make a table and so far you are given the information below:

	Lung cancer	Healthy	Total
Radon in home	a =	b =	a+b =
No Radon in home	c =	d = 500	c+d = 550
Total	100		1000

Based on this we can fill this in further:

	Lung cancer	Healthy	Total
Radon in home	a = 50	b = 400	a+b = 450
No Radon in home	c = 50	d = 500	c+d = 550
Total	100	900	1000

The odds ratio is calculated this way

$$OR = \frac{a \cdot d}{c \cdot b}$$

$$OR = \frac{50 \cdot 500}{50 \cdot 400}$$

$$OR = \frac{5}{4} \ or \ 1.25$$

See page 43 for more information.

6) **Answer d)** If you remember from page 49 the reduction of bias can be accomplished by **blinding, adequate randomization, control groups,** and through **crossover studies (control for confounding variables)**. The case control study was random and had a control group by definition, but nothing was said about blinding. Remember that triple blinding means that the patients and the researchers do not know who is in which group plus the statisticians analyzing the data.

7) The answers are in bold:

Example	Bias
i) A study is performed to see the effects of a new class of lipid lowering medication. A trial is conducted on inpatients in hospitals across the country.	a) Berkson
ii) A survey is sent out to those who have Huntington's disease. The average age of those taking the study is 41.	b) Prevalence
iii) A set of surveys is sent out a population asking about antibiotic use habits such as if they finish the whole prescription as asked. 97% say yes. It is later revealed only healthcare providers took the survey.	c) Non-response
iv) A new form of chemotherapy is set for trial. The groups are randomized and blinded. The medication seemed effective, but many people in the treatment group end up dropping out due to the extreme side effects of the medication.	d) Attrition
v) A 16 year old girl does a science project on antibiotic use among households, but only asks people living in a more affluent neighbourhood.	e) Ascertainment

vi) To see the how intravenous drug use relates to income status several surveys were sent out to randomized strata of different income levels. It is later discovered that many people in more affluent communities did not tell the truth in fear of how others would think if they found out.

f) Reporting

vii) To discover adverse effects from a new version of acetaminophen households to see if they can remember adverse outcomes versus the benefits. People who experienced some mild abdominal discomfort were less likely than those who did not have adverse effects to report that the medication relieved their headache.

g) Recall

viii) During a trial looking at Med A and Med B the physicians have heard that Med A causes anemia more often than Med B. The doctors make sure to run CBCs more often on people with the Med A group. They study finds Med A does cause anemia.

h) Surveillance

ix) During a study researchers discover there is a significant difference between two groups. It is found that each researcher used their own criteria for a certain diagnosis.

i) Observer bias

x) A particular cancer has a mean survival of 5 years. A new test discovers the cancer six months earlier and with it changes the survival to 5.5 years.

j) Lead time

xi) A physician runs a study where he does not put in adequate controls to ensure adherence in the placebo group.

k) Procedure

xii) A nutritional study is performed to see how dietary habits affect mood. The physician sees those in the study daily and asks about what they have been eating.

l) Hawthorne

8) **Answer c) sensitivity decreases, PPV increases**. This is just a little curveball question using Chapter 2 knowledge. It is helpful to look at the following graphics.

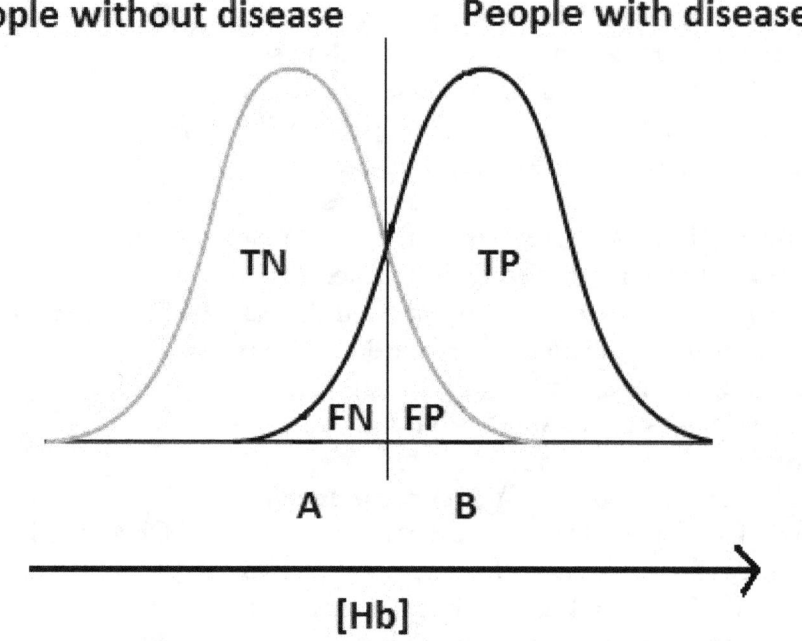

This is the normal spread of where TN, TP, FN and FP fall.

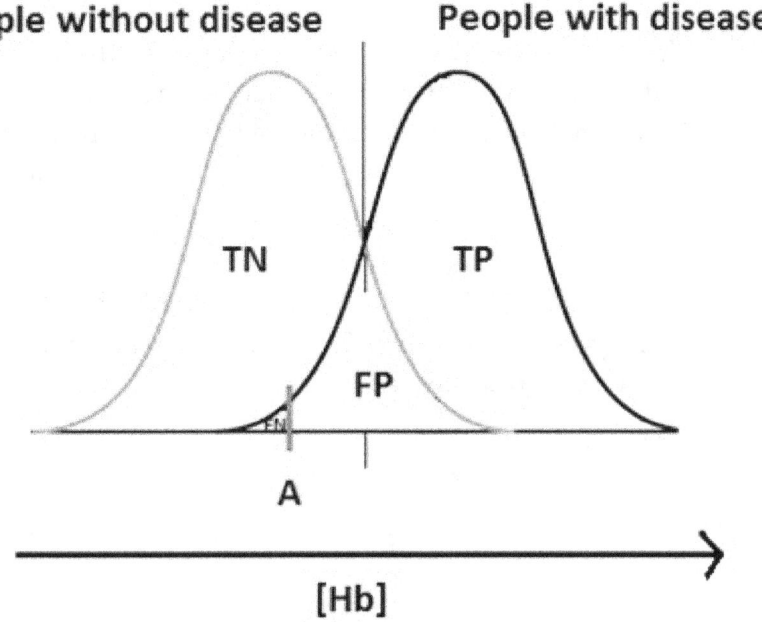

With the cut off at A the FP (false positives) predominate, with few false negatives. Bellow you will see what happens if we change the cut off to B.

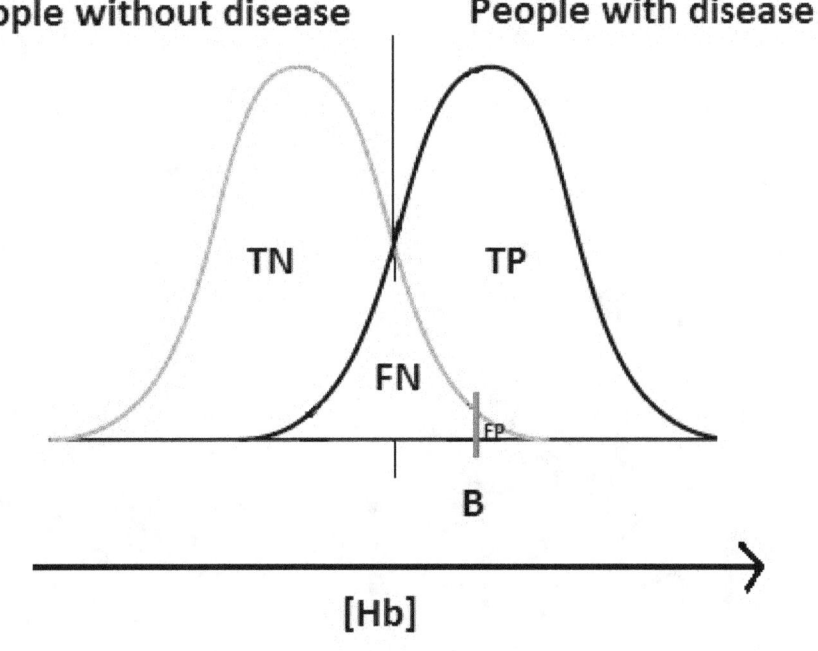

You can see that by changing the cutoff to B you create more FN, and decrease the FP. If you remember from the equations, FP and FN are in the denominator of all of them. That means if you decrease one it will increase the overall value and vice versa.

$$Sensitivity = \frac{TP}{FN + TP}$$

$Sensitivity = \frac{6}{4+6} < Sensitivity = \frac{6}{2+6}$ for example. Applying this concept to all gives us a **decrease in sensitivity and NPV** and an **increase in specificity and PPV.**

Chapter 4: Handling USMLE Study Questions

The final chapter of this book will consist of two example problems. The exam may contain a study that you have to answer two to three questions on. The objective is just to see if you can interpret the results of a basic study. Try to solve the questions associated with these two **fictitious drug/health studies**.

Practice Study 1

A small serving of sucrose before meals encourages weight loss

Methods

Design: Randomized clinical trial
Blinding: Patients, clinicians, and statisticians blinded
Follow-up period: 12 months
Setting: Outpatient
Patients: 9 patients were selected (mean age 34, 55% male) out of a group of patients who wanted to lose weight. Each group was given instructions for how to lose weight through diet and exercise. For uniformity, each was placed on the DASH diet. Exclusion criteria were history of difficult to control diabetes mellitus, major planned surgeries in next 6 months, and pregnancy.
Intervention: The experimental group was given instructions to use the 1 gram sucrose (table sugar) packets provided before each major meal (breakfast, lunch, dinner). Regular monthly meetings occurred to check adherence, with every third month checking the A1C.
Outcomes: Increase in weight loss and waist size in the experimental group.

Patient Follow Up: 98%

Results

At the end of 12 months the experimental group had an overall decrease in weight and waist size, but no change in the A1C.

Sucrose vs Control Group for Weight Loss (Male)							
	Sucrose			Control			
	Start	One year	Change	Start	One year	Change	P value
Change in weight (Kg)	79.3	74.3	5.0	79.1	74.8	4.3	0.03
Change in waist size (cm)	102	93	9	103	95	8	0.02
Change in A1C (%)	6.35	6.25	0.10	6.34	6.23	0.11	0.11

Sucrose vs Control Group for Weight Loss (Female)							
	Sucrose			Control			
	Start	One year	Change	Start	One year	Change	P value
Change in weight (Kg)	67.0	61.0	6.0	67.3	62.2	5.1	0.01
Change in waist size (cm)	90	84	6	90	85	5	0.03
Change in A1C (%)	6.36	6.27	0.09	6.31	6.11	0.20	0.09

Conclusion

In patients attempting to lose weight taking 1 gram of sucrose before major meals will increase weight loss and waist size.

1) Based on the above results you can conclude which of the following?
    a. Statistically significant and clinically significant

      b. Statistically insignificant and clinically insignificant

      c. Statistically significant, but clinically insignificant

      d. Statistically insignificant, but clinically significant

2) A 32 year old female is in your office asking about a weight loss article she read on the internet. You look up the study she refers to and it is the one above. Which of the following can be said about the flaws of the study?

      a. By P values we can conclude the results are unlikely to suffer from a type I error

      b. The power of the study is likely high

      c. By P values we can conclude the results are unlikely to suffer from a type II error

      d. The probability of type I error is high

      e. Too short of a follow up was used

Practice Study 1 Answers

1) **Answer c)** You see the results are statistically significant but the differences in weight loss is a kilogram or two from the placebo. This means in practice (ie. clinically) the results are not very significant.

2) **Answer d)** It is important to remember the number of participants was 9. The **power** of a study is related to the number of participants and is directly related to the **probability of making a Type I error.** Remember a **P value does not correlate to the probability of making a Type I error** (rejecting the null hypothesis when it is true).

Practice Study 2

A new anticoagulant for use in those with atrial fibrillation, called Kind Of Works (KOW) is compared against warfarin. It should be noted that at this time KOW does not have a reversal agent.

Methods

Design: Randomized clinical trial
Blinding: Patients blinded, clinicians non-blinded

Follow-up period: 3 years

Setting: Outpatient

Patients: 143 patients were selected (mean age 69, 43% male) out of a group of patients who required long term anticoagulation. Exclusion criteria included a major bleed in the last three months, expecting surgery in the next six months, and life expectancy less than one year. Anyone who had a major bleed was required to stop whichever medication they were taking.

The results of the study are shown below.

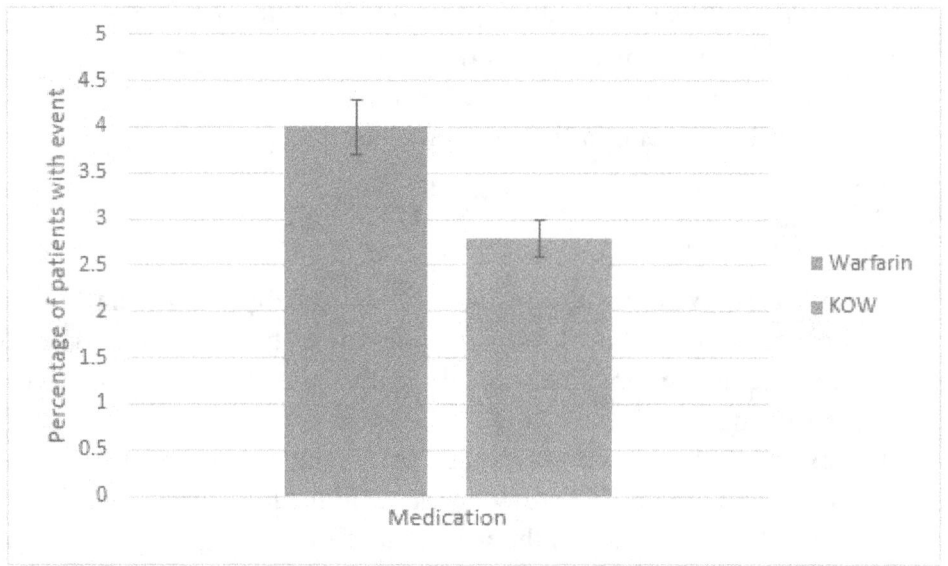

Warfarin = 4% +/- 0.3%

KOW = 2.8% +/- 0.2%

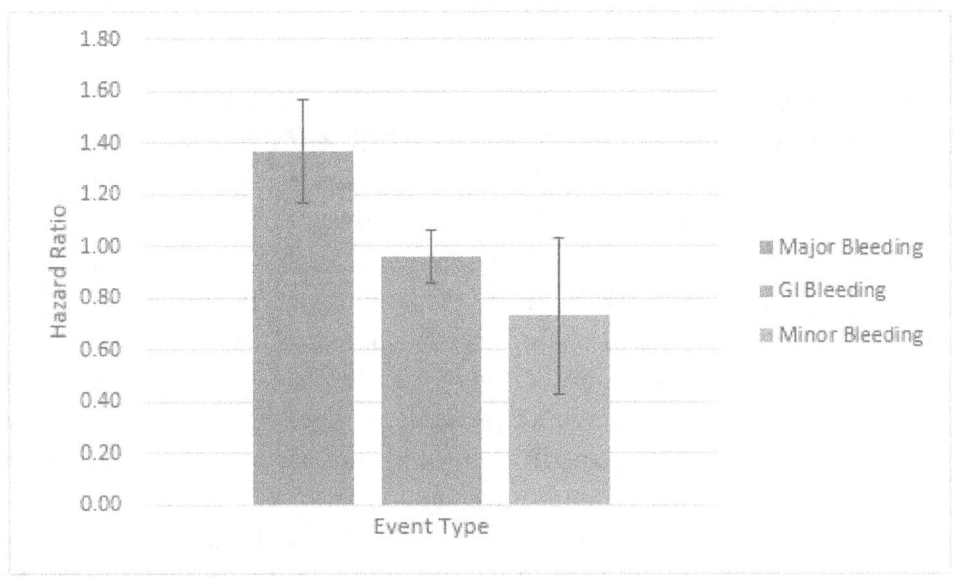

HR Major Bleeds = 1.37 +/- 0.2
HR GI bleeding = 0.96 +/- 0.1
HR Minor Bleeds = 0.73 +/- 0.3

1) Based on the results which of the following conclusions can be drawn?
    a. KOW has less major bleeding events compared to warfarin
    b. The decreased minor bleeding is statistically insignificant
    c. The NNT to prevent stroke with KOW vs Warfarin is 2
    d. KOW has fewer GI bleeds than warfarin
    e. Warfarin should be used because it has a reversal agent
2) A further study from the drug company shows that warfarin drops Hb levels by 1.1 g/dL in men on average while KOW drops it by 1.5 g/dL on average. Based on this a concerned son says he wants his father on Warfarin to make sure his Hb stays higher. You decide that is not the best reason to decide one over the other. What do you think made the physician come to this conclusion?
    a. Sample size of the study is too small
    b. Duration of follow up too short
    c. The Hb outcomes are not important
    d. Exclusion criteria was not strict enough
    e. KOW is still superior because it decreases the number of strokes

Practice Study 2 Answers

1) **Answer b)** The key point to recognize is that in terms of a **HR it is a ratio of adverse events in the treatment group to the control group.** This means a **value > 1 represents a greater risk** in the **treatment group**. In terms of a standard error if the interval contains 1 this means the results are insignificant. In this case the greater risk of GI and minor bleeding in KOW are not significant, but KOW has a significant risk of major bleeds compared to Warfarin. Unlike what **choice c)** suggests the **NNT is 67** (1/(0.04-0.028)). Just because a "reversal agent" is available does not make it superior. Warfarin has many medication/food interactions and the antidote does not improve outcomes in major bleeds.

2) **Answer c)** Always remember to treat the patient and not the numbers. If a particular value changes (Hb, potassium, LV function) you must discover if this is significant or not. The change in Hb is so small it is not likely to create a significant anemia; however, you can still check to see if it does. Since there is a **10% chance of stroke each year in patients with A-fib** a study that lasts 3 years is an adequate amount of time to watch the patients, so **choice b)** is out. From question 1) you realized the NNT is 67 and it has an increased risk of major bleeds compared to Warfarin, so **choice e)** does not fit. The exclusion criteria was also adequate and bias free.

A final note on study designs. You may hear the term intention to treat which just means that any drop outs from the study are included. This reduces bias from exclusions, but can lead to a loss of statistical power if you do not adjust your sample size for this outcome. This method can also cause bias if the patients from the study were removed due to protocol violation.

You have reached the end of the book! Thank you for reading and again I can be reached at mccee.tutor@gmail.com or through my website at http://www.mcceetutoringservices.com/contact for more information on how my service can help prepare you for your exams!

Appendix A – Percentiles

Percentile	Z-score	Percentile	Z-score	Percentile	Z-score
1	-2.326	18	-0.915	35	-0.385
2	-2.054	19	-0.878	36	-0.358
3	-1.881	20	-0.842	37	-0.332
4	-1.751	21	-0.806	38	-0.305
5	-1.645	22	-0.772	39	-0.279
6	-1.555	23	-0.739	40	-0.253
7	-1.476	24	-0.706	41	-0.228
8	-1.405	25	-0.674	42	-0.202
9	-1.341	26	-0.643	43	-0.176
10	-1.282	27	-0.613	44	-0.151
11	-1.227	28	-0.583	45	-0.126
12	-1.175	29	-0.553	46	-0.100
13	-1.126	30	-0.524	47	-0.075
14	-1.080	31	-0.496	48	-0.050
15	-1.036	32	-0.468	49	-0.025
16	-0.994	33	-0.440	50	0
17	-0.954	34	-0.412	51	0.025

<u>Appendix B – Nomogram</u>

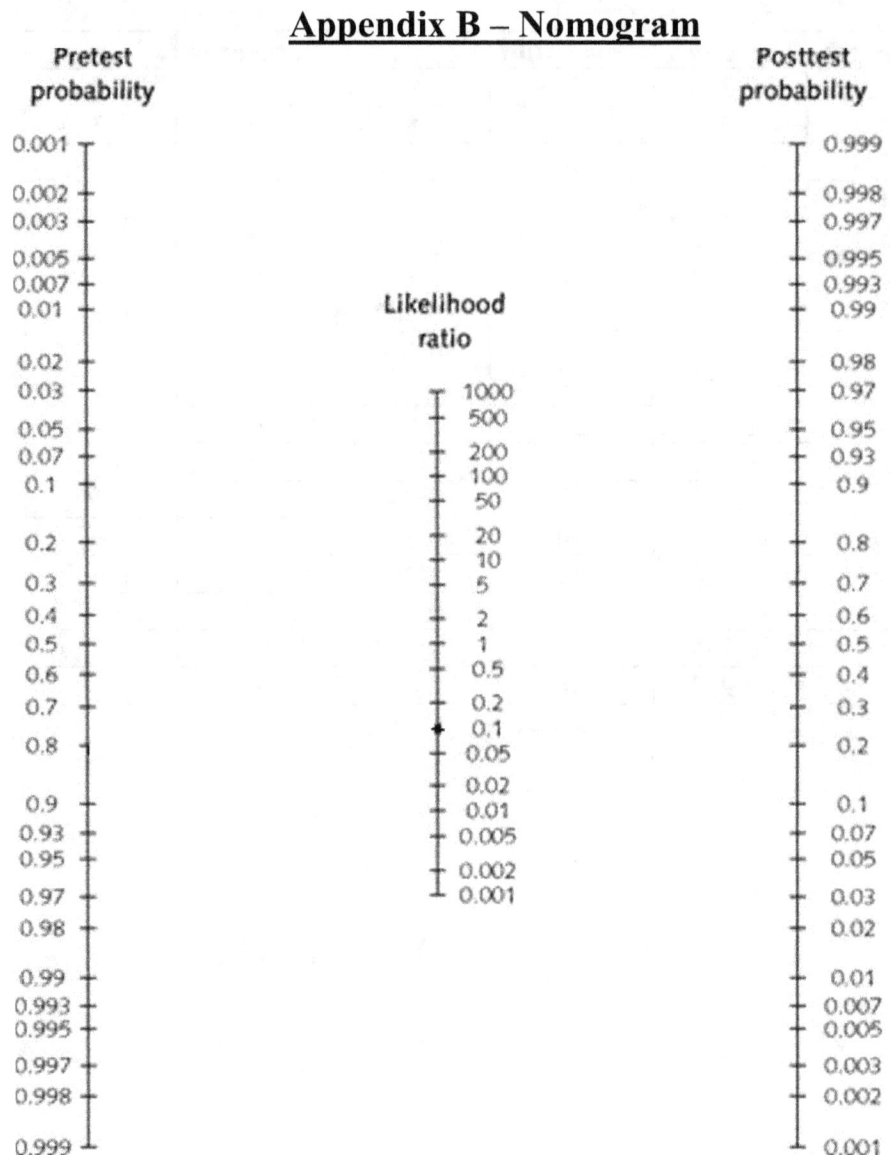

Index

References

Gordis, L. (2009). *Epidemiology Fourth Edition.* Philadelphia, PA, USA: Elselvier Inc.

Hoffman, R. (2016, July 13). *Screening for prostate cancer*. Retrieved from UpToDate: https://www.uptodate.com/contents/screening-for-prostate-cancer?source=search_result&search=prostate%20cancer%20screening&selectedTitle=1~62

Le, T., Bhushan, V., Kulkarni, V., & Sochat, M. (2013). *First Aid for the USMLE Step 1 2013.* McGraw Hill.

Muller, D. (2016, August 11). Is Most Published Research Wrong? Veritasium.

Reinhart, A. (2015). *Statistics Done Wrong: The Woefully Complete Guide.* William Pollock.

Strogatz, S. (2012). *The Joy of X: A Guided Tour of Math From One to Infinity.* New York, NY, USA: Houghton Mifflin Harcourt Publishing Company.

Notes

Notes

Notes

Notes

Notes

<u>**Notes**</u>

<u>Notes</u>

Medical School Statistics Rx: Statistics for the MCCEE and USMLE

Notes

Notes

Notes

Notes

Notes

Notes

Notes

Notes

Notes

Medical School Statistics Rx: Statistics for the MCCEE and USMLE

Notes

Notes

Notes

www.ingramcontent.com/pod-product-compliance
Lightning Source LLC
Chambersburg PA
CBHW081508170526
45166CB00008B/2588